Auri legendi ratio in rivis è montibus Apa-   XLI.
latcy decurrentibus.

ROCVL ab eo loco, in quo nostra arx extructa fuit, magni sunt montes, Indorum lingua, Apalatcy cognominati, e quibus, ut ex topographica charta videre licet, oriuntur tres magni rivi, provolventes arenam, cui multum auri, argenti & æris admixtum est. Eam ob causam illius regionis incolæ fossas in rivis faciunt, ut provoluta ab aqua arena, in eas propter gravitatem cadat: diligenter inde eductá in certum locum deferunt, & aliquanto post tempore, denuo fossis arenam quæ incidit exhaurientes, colligunt, & cymbu imposuam per ingens flumen devehunt, à nostris Maii nomine insignitum, quod in mare sese exonerat. Opes quæ inde proveniunt, nunc Hispani in suum usum convertere norunt.

PLACER MINING BY INDIANS OF GEORGIA.

SMITHSONIAN INSTITUTION

BUREAU OF AMERICAN ETHNOLOGY

The Mining of Gems and Ornamental Stones by American Indians

By SYDNEY H. BALL

Anthropological Papers, No. 13

From Bureau of American Ethnology BULLETIN 128, pp. xi–77

UNITED STATES

GOVERNMENT PRINTING OFFICE

WASHINGTON : 1941

SMITHSONIAN INSTITUTION
Bureau of American Ethnology
Bulletin No. 128

Anthropological Papers, No. 13

The Mining of Gems and Ornamental Stones by American Indians

By SYDNEY H. BALL

CONTENTS

ILLUSTRATIONS

PLATES

THE MINING OF GEMS AND ORNAMENTAL STONES BY AMERICAN INDIANS

BY SYDNEY H. BALL

*I tell thee, golde is more plentifull there [Virginia] then copper is with us * * * Why, man, all their dripping-pans * * * are pure golde: * * * and for rubies and diamonds, they goe forth on holydayes and gather 'hem by the sea-shore, to hang on their childrens coates, and sticke in their childrens caps, as commonly as our children weare saffron-gilt brooches and groates with hoales in 'hem.*[1]

INTRODUCTION

When Europeans arrived in America they found the American Indian largely in the Stone Age, although a number of tribes, and particularly those of Mexico, Central America, Colombia, and Peru, used certain metals. Imbued with our conception of racial superiority, we rarely think of the Indian as a capable prospector and a patient, if primitive, miner. Yet the rapid development of mining in Mexico and Peru after the conquests was mainly owing to the large number of ore bodies opened up by the local aborigines. Lust on the part of the Spaniards for gold, silver, and precious stones and, to express it mildly, canny concern on the part of the English and French for such wealth, were the activating forces behind much of the exploration of America. Both to the American Indian and the white man, mineral products were essential, but the former used coal mainly as an ornament and petroleum as a liniment, while the latter could not be inveigled to "rush" a new obsidian "find" to supply weapons of war.

This article treats of the gems and ornamental stones used by the Indian before he came in contact with the white man. His metal mining has been frequently described: The pits he dug on the Lake Superior Copper Range; his exploitation of the mercury mines of New Almaden and Peru for paint; his placer mining in Georgia, Mexico, Central America, the West Indies, northwestern South America, and Brazil; and his gold and silver lode mining in Mexico

---

[1] Seagull, a sailor, to Scapethrift in Eastward Hoe, a popular drama by Geo. Chapman, Ben Jonson, and Joh. Marston, 1605. See volume 3, p. 51, of The works of John Marston, 3 vols., London, 1856.

1

and Peru. Copper and gold were extensively used; silver, tin, lead, platinum, mercury, and iron (meteoric) less so. Within the United States copper and gold were treated as pebbles and pounded into the shape desired, but in Mexico, Central America, and western South America an elementary smelting technique had been evolved and in certain localities the skill with which metals were forged, cast, alloyed, and plated astonished the conquistadoies. Platinum, it may be recalled, was used by the Indian long before it was known to white men.

The Indian's knowledge of gems and ornamental stones was, however, much more comprehensive than that of metals and his most extensive mines or, at least, those known to us, were of ornamental stones.

The Indian made use of a large number of gems and ornamental stones, some 84 being known to the writer. His acquaintance with minerals suitable for decorative purposes exceeded in number, at least, that of the peoples of Europe and Asia at the time of the discovery of America. The list, as presented (p. 56) is fairly complete for North and Central America but for South America could doubtless be appreciably extended. Of the early men who made the Folsom points, those of Folsom, N. Mex., used chalcedony, jasper, and obsidian, and those of the Lindenmeier site near Fort Collins, Colo., chalcedony, jasper, moss agate, lignite, quartz, hematite, and agatized wood. It can be said with considerable assurance that at the beginning of our era, or roughly 2,000 years ago, the American Indian used in addition the following precious and decorative stones: Agate, alabaster, azurite, bloodstone, calcite, jadeite, jet, lapis lazuli, malachite, mica, nephrite, common opal, pyrite, satin spar selenite, serpentine, soapstone, and turquoise. By 1000 A. D. or even earlier, the following, among others, had been added: Amber, carnelian, catlinite, chloromelanite, emerald, fluorspar, galena, garnet, magnesite, marcasite, moonstone, noble opal, sodalite, and variscite.

It is a curious instance of parallel cultural growth that of the 25 stones first used by early man on the eastern and western continents, 16 were common to the two cultures (Ball, 1931, pp. 683–685).

As sources of material on the subject, we have: (1) Artifacts preserved in public and private collections; (2) mine workings, although many which once existed have become obliterated; (3) early and present-day literature; and (4) the traditions and myths of the Indians. Evidence from the latter two sources must be used with discretion, as the mineralogic knowledge particularly of the early Spanish writers, and of some archeologists, is, to say the least, inadequate, while myths may be of much later origin than the events they describe.

USES OF GEMS AND ORNAMENTAL STONES BY AMERICAN INDIANS

Gems and ornamental stones were used by the aborigines for a large number of different purposes, for nonmetallic minerals served not only their special functions as we know them but also, at least among certain of the Indian tribes, all the uses of metals and, among all of them, certain of the functions of our metals. The principal uses may be listed as follows:

a. Ornaments.—Ornaments consisted of pendants, beads, and carved figures: Gem mosaics were worn among the Pueblos, the semicivilized Mexicans, the Peruvian peoples, etc. The Indian adores vivid colors and he has a childlike love of beautiful pebbles, particularly if brightly colored. To the semicivilized peoples from Arizona to northern Chile, the blue and the bluish-green of the turquoise and the green of jade and emerald had a peculiar fascination. Such stones were eagerly sought and highly valued. The first gained its color from the cloudless sky and the second symbolized the growth of crops; their value being increased by the supernatural power gained thereby. So highly regarded were they that jade could be worn only by nobles among the Aztecs, and among the Yavapai (central Arizona) only a chief (*mastova*) was privileged to wear turquoise bracelets (Gifford, 1932, p. 229). All green stones were treasured by the Yuma medicine men (Corbusier, 1886). The Indian used many materials in their jewelry which, until the introduction of the modern "novelty" jewelry, we would have scorned. Shell, for example, and wood, berries, seeds, iridescent beetle wings and fruit stones, and even worse, for Capt. John Smith, in his history of Virginia, tells us that some of the Virginia Indians wore in holes in their ears small "green and yellow colored snakes near one-half a yard in length"; others "a dead rat tied by the tail." The Cheyenne Indians strung human fingers into necklaces (Bourke, 1890), and the Sioux wore hands as earrings (Mallery, 1893, p. 752, fig. 1278). On the other hand, the custom of the women of the southern Mexican coast and of British Honduras of placing fireflies in their hair, while a bit startling to the white traveler, has, after all, charm.

The use of jade, hematite, turquoise, rock crystal, obsidian, and pyrite inlays in the teeth of Mayan and certain Mexican and South American tribes was, of course, for display rather than as an aseptic measure.

b. Weapons.—Hard stones with conchoidal fracture were eagerly sought for arrow- and spear-heads, knives, and razors. That some of these weapons were effective, we learn from rough old Bernal Diaz, who asserts that a single blow of an Aztec sword, set with obsidian points, would decapitate a horse. The Hudson Bay Eskimo, when lead is scarce, use soapstone as bullets.

c. Household utensils.—Soapstone was widely used for lamps, particularly among the Eskimo, and for cooking utensils by many tribes. Pyrite and quartz were used to produce fire, particularly by the peoples of the extreme northern part of the American Continent (the Eskimo) and of the extreme southern part (the Fuegians). Angular fragments of quartz, and some say obsidian, set in a flat board, are made by a British Guiana tribe and are traded over considerable distances as manioc graters (Farabee, 1917, p. 77). Commerce in these graters was widespread (McGovern, 1927, p. 211). Similar graters were used by the Uaupes in the upper Amazon (Wallace, 1853, pp. 483–484).

d. Surgical instruments.—Shaped rock crystal was used for blood-letting by the California Indians, and lancets of this material and obsidian by the ancient people of Peru. Obsidian knives served their grewsome role when victims were sacrificed to the gods of Mexico, and knives of this material were used in scarification by California Indians and by the **Araucanian** and the Mapuchian shamans in bleeding the sick.

e. Graving tools.—The Hidatsa of North Dakota cut their pictographs on rock with sharp-pointed quartz fragments, as Richard Spruce infers the Amazonian Indians did.

f. Abrasives.—The Indians of Racine County, Wis., crushed rock crystal to form an abrasive in pipe making (West, 1934, p. 341). Pumice was also used as an abrasive by the Nevada, Nebraska, Montana, and California Indians and by the Eskimo of Cook Inlet, Alaska.

g. Mirrors.—Mirrors were made of obsidian and pyrite by the Aztecs, Mayas, and the Peruvians; and among the Mound Builders mica probably so served. The Eskimo of Hudson Strait use a plate of biotite "so fitted into a leather case as to be seen on either side" (Lyon, 1825, p. 38).

h. Windows.—Selenite and perhaps mica were used in windows by the Pueblos, and Mexican "onyx" by the Aztecs. Selenite was supposed to permit the Pueblos to see out but the keenest eye could not see what was passing in the interior of the feebly lighted rooms (Möllhausen, 1858, vol. 1, p. 157).

i. Embellishment of buildings.—Precious stones were used in quantity by both the Aztecs and the Peruvians in their temples and palaces. Turquoise was, in instances, set in the lintels of the Pueblo houses. Arizona agatized wood was sometimes used as a building stone in constructing the ancient pueblos.

j. Pigments.—Hematite, malachite, and azurite were not only widely used as pendants and in other ways in the mass but, when crushed, as pigments. The latter two furnished the Pueblo people their favorite colors—green and blue. Among the Navaho crushed turquoise was used to paint certain ceremonial objects (Pogue, 1915, p. 103). The green used to dye the wool of Chilkat blankets was derived from copper ores. The British Columbia Indians also used malachite as a pigment. The Pawnees and Mandans heated selenite and from the powder made a whitening used in tanning buckskin. The Navaho medicine men used gypsum as chalk in drawing and the Pueblos merely powdered, or burnt and mixed it with water as whitewash. Calcined gypsum powder was used by the Omahas to clean, whiten, and dry the sinews binding feathers to their arrows. The Aztecs used *chimaltizatl* (selenite) to whiten their paintings (Clavigero, 1807, pp. 16–17). The California Indians procured body paint from a "vermilion cave," the outcrop of the New Almaden mercury mine. Cinnabar was also used by the Aztecs, the Mayas, and the Peruvian Indians. The beautiful pale green brochantite of the Corocoro copper deposit, Bolivia, was used by the local Indians as a source of green pigment before the Spanish arrived (Berton, 1936). The Indians used the brilliant red hewettite (a hydrous calcium vanadate) to make pictographs on the sandstone cliffs of Emery County, Utah. Within one-half mile are commercial vanadium deposits.[2] Black pigments were produced from lignite (Pueblos), from manganese dioxide (Pueblos and Californians), from coal (Haidas), from graphite (New York, New England, and Alaska Indians and Eskimo), from sphalerite ore (Pueblos), from micaceous hematite (Yukon Indians), or from galena (Apache-Yumas). The latter also used calcite and magnesite as white pigments. The Oubeways, on the other hand, used iron sulphate derived from decomposing pyrite as a black dye. The Seri, inhabiting Tiburon Island, Sonoro, used dumortierite as a blue face paint (Kroeber, 1931, p. 27). The Pueblos used jarosite in addition to yellow ochre for yellows and browns (Cosgrove, 1932).

k. Currency.—Among the Indians of California, large obsidian blades and beads made of magnesite were used as standards of barter. The larger of the

---

[2] Written communication from Robert H. Sayre, Denver, Colo.

latter were worth up to $20 (Hodge, 1922, p. 16; Sumner, 1907, p. 152). Black obsidian blades of ordinary length were worth among the Yurok Indians of northern California about $1 an inch; the red obsidian, being rarer, was even more valuable. Blades of 30 to 33 inches in length were worth much more, indeed they were of inestimable value (Kroeber, 1925, pp. 26–27). Turquoise was, perhaps, as near to a unit of value as the Pueblos, Navahos, and Yaquis possessed and nephrite beads were used in somewhat the same way in British Guiana and in the Mayan cities.

Like most primitive people, the American Indian saw in gems and decorative stones not only beauty but the supernatural and the awe-inspiring. The medicine men among the Cayapa Indians of Ecuador, for example, are unwilling to dispose of the pebbles used in their incantations lest the spirit in the stone become angry, an old Greek idea. Edward Bancroft (1769, pp. 21, 311) states that the Indians of British Guiana in his time would not touch agate pebbles locally abundant "from a principle of superstitious veneration, as they are dedicated to the service of their magical invocations." Precious stones, therefore, were important factors in their religious life as the following uses indicate:

l. Objects of worship.—Among the pre-Colombian aborigines of Manta, Ecuador, a large emerald was worshipped, and rock crystal and jasper appear to have served the same purpose, respectively, among certain of the Peruvian and Ecuadorian tribes, until the Incan army forcibly showed them the error of their ways and they became sun worshippers. The Mixtecs worshipped a large jade at Achiuhtla, representing the god Quetzalcoatl. There is now at the University of Texas a meteorite from Wichita County, Tex., long venerated by the Comanche for extraordinary curative powers. It was known as *Po-a-cat-le-pe-le-corre* (Medicine Rock). In passing, all made votive offerings to it (Bolton, 1914, vol. 1, p. 296).

m. Fetishes and charms.—Rock crystal among the Natchez (Le Petit *in* French, 1851, pt. 3, p. 141) and the Pimas, nephrite among the Brazilian fishermen, sections of *Baculites* among the Cheyennes, turquoise among the Pueblos, and malachite among the Apaches, served as fetishes. Quartz crystals were used as charms by the Eskimo shamans, as was jasper, and the medicine men of the Tarahumara today use quartz crystals as charms (Bennett and Zingg, 1935, p. 369). Among the Yumas quartz crystals brought good luck.

n. Votive offerings to gods.—Turquoise among the Pueblos and Navahos, emeralds among the Chibchas of Colombia, emeralds and turquoise among the Incan Indians, and precious stones among the Tahus of Sinaloa, served as votive offerings. The Zuñi women ground corn and mixed it with powdered turquoise and white shell and offered it to their gods as food.

o. Temple incense.—Amber was used as incense by the Aztecs.

p. Means of divination.—Quartz crystal among the Pueblos and the Cherokees (Olbrechs, 1930, pp. 547–552), hematite among the latter tribe, obsidian and pyrite mirrors among the Aztecs and the Cakchiquels of Guatemala, and jade globes and rock crystals (Brinton, 1883, p. 245) among the Mayas, were used in divination. The Navaho medicine men use rock crystal in diagnosing disease by star-gazing, the light of the star being reflected in the stone permitting him to see the cause of the sickness of the patient like a motion picture (Wyman, 1936, p. 245).

q. Medicinal use.—Gypsum, ochres, and other minerals were used medicinally. Among the Tehuelche, a Patagonian tribe, "the new born babe is smeared over with damp gypsum" (Musters, 1871, p. 176), doubtless a reasonably good substitute for talcum.   Powdered steatite was similarly used on Washoe and Yakut babes.   Locally among the Colombian Chibchas the stomach of the desiccated corpse of the chief was filled with gold and emeralds before the body was wrapped in cotton and placed on a pedestal in a corner of one of their shrines.   (Perhaps not medical, but interesting.)

r. Musical instruments.—Like the Chinese, the Venezuelan Indians knew the sonorous tones when thin plaques of nephrite are struck (Heger, 1925, pp. 148–155). The Pomo Indians of California also recognized the sound-producing qualities of minerals for they suspended two blocks of obsidian in the tree above which their deer traps were fastened.   The struggling deer caused the stones to clash together, thus appraising the trapper of his success.

KNOWN SOURCES OF GEM SUPPLY

The maps of North and South America (pls. 4, 5) and list of mines operated by the Indians (p. 56) show the sources known to the writer from which the Indians obtained gems and ornamental stones.   The list is markedly fragmentary for a number of reasons; principally because the mines, quarries, and placer deposits, abandoned for the most part centuries ago, have disappeared or become inconspicuous and also because much of the product sought by the Indians was of little or no value to the white.   For the latter reason the list of pre-Hispanic quartz mines is doubtless much less complete than are the lists, for example, of emerald and turquoise.   However, in spite of these difficulties, 289 Indian mines are listed.

Within the writer's knowledge of the archeology of the various areas and, taking into consideration the extensive territory occupied by the tribes and the relative mineral wealth thereof, it may be stated that the tribes with the widest knowledge of decorative stones and gems were the Mayas, the Pueblos, the Aztecs, the Mound Builders and the Indians of our southeastern States; a list including most of the more advanced peoples who were not only more ingenious than the average savage but whose higher civilization demanded a greater variety of raw materials.   As miners, the Pueblos, Aztecs, Peruvians, and the Indians of the southern Appalachians were outstanding while the Mayas and Mound Builders depended largely on commerce for their supplies of gems.

THE INDIAN AS GEOLOGIST AND MINERALOGIST

The Indian, generally using tools of stone, by long experience became a fair geologist, knowing the rocks most suited to his needs and their characteristics—indeed, probably much better than we do.   This required a knowledge of texture, hardness, and mineralogic make-up, so that he could recognize the same mineral if found in two different places.   One mineral was good for pipes, a second for arrow points, and still a third for axes.   The Indian's curiosity regarding minerals is shown by the presence in graves of mineral fragments apparently not valuable to him economically, such as zinc blende and chalcopyrite (Schenck and Dawson, 1929).   Incidentally, the Indian mineralogist knew atacamite, sodalite, brochantite, pectolite, labradorite, dumortierite, hewettite, and variscite long before his European confrere.

One reason the Indian so frequently sought material for his arrowheads from stream gravel was that he realized that such material was fresh and did not shatter badly, as opposed to weathered material from surface outcrops or detrital boulders.   Capt. G. F. Lyon (1825, p. 69), on his visit to Southampton Island, found both flint and agate pebbles on the beach, but the Eskimo use only the former, since the latter are prone to split along the banding.   To obtain absolutely fresh arrow material, whether flint or obsidian, the Indian was at times forced to quarry, and, in quarrying, the partially weathered surface material was rejected.   Similarly, the makers of catlinite pipes rejected all but a small part of the material quarried, for it had to be heat-resistant, of good color, and easily shaped.   The Indian realized that soapstone absorbs and retains heat and that lamps or cooking utensils of it, once heated, gave up their heat slowly, keeping the oil fluid or the game cooking.   The California Indians used the softer, more micaceous Santa Catalina steatite for pots and the closer-grained darker rock for weights of digging sticks, pipes, and ornaments (Schumacher, 1880, pp. 259–264).   As the arrow-shaft straighteners were first heated, the Indian had to select for this use a mineral or rock resisting heat; soapstone and serpentine were common materials.

The Pomo Indians of California anticipated the modern petrographer's ideas as to quartz by differentiating obsidian into two distinct types, *bati xaga* ("arrow" obsidian), which was especially suitable for flaking, and *dupa xaga* ("to-cut" obsidian), which was harder, broke more cleanly, and was consequently, for instance, suitable for razors.

The first came from Lower Clear Lake, the second from Cole Creek (Loeb, 1926, p. 179).

The luminescence of rock crystal (q. v.) was known to the Pueblo Indians and it is not impossible some arrow maker of our own Southwest many centuries ago was the first to observe this property of minerals.

IDEAS AS TO ORIGIN OF PRECIOUS AND DECORATIVE STONES

The Indian normally accepted minerals unthinkingly, but in some instances an aboriginal philosopher tried to explain their origin.   His results were reminiscent of those of his Greek prototype: Minerals were the direct gift of the Great Spirit (the Sauk and Fox idea regarding the galena of the Upper Mississippi Valley) or their origin was suggested by some striking physical property. Certain of the Indians of the north country saw in rock crystal a form of ice, as did the ancient Greeks and as do the modern Afghans; to the Mohawks, however, its glistening crystals were the congealed tears of a devoted mother, reminding us of a Greek myth as to the origin of amber. Sparks were derived from pyrite, therefore to the Point Barrow Eskimo it fell from the sky. Laminae of mica to the Delaware were scales of the mythical Horned Serpent. Catlinite from its color evidently was either stained by buffalo blood or was hardened human flesh. Again flint was associated with war and the chase; consequently to the Potawotomi, flint pebbles found here and there mark the sites of combat between an evil spirit and Nonaboojoo, "the friend of the human race" (De Smet, 1847; Thwaites, vol. 29, 1904–7, p. 376). Similarly, the silicified wood found on the mesas near the Grand Canyon, Ariz., was known to the Indians as the arrows of Shin-ar'-ump (Powell, 1875, p. 190). The brittleness of obsidian accounts for the Pomo Indians' explanation of the abundant fragments of the rock on Mt. Kanaktai, north of San Francisco Bay, Calif.; namely, that the obsidian-man, caught in a bush, in freeing himself fell and broke into thousands of pieces (Barrett, 1933, vol. 15, pp. 31, 220, 226, 231).

In instances the Indian was poetical, as is well exemplified in his legend of the origin of the iridescent obsidian which occurs with the ordinary obsidian at Glass Butte, Lake County, Oreg. To kill off a particularly venomous type of yellow jackets, the Indians surrounded the mountain and set fire to the forest.

After the mountain had burned for several days there came up a shower having a beautiful rainbow. The rainbow shone all day on one spot on the south side of the Mountain and at evening seemed to enter the ground at this particular spot. After the fire they found that some of the mountain had melted and had made heaps of glass for their arrow and spear points, but the rainbow had settled into one heap and left the beautiful colors there.

This they called "rainbow" obsidian and used only for sacred and religious purposes (Forbes, 1935, pp. 307–309).

On the west shore of Lake Champlain near the falls, the waves throw well-polished flint pebbles upon the shore. The Mohawk companions of Fathers Pierron, Fremin, and Bruyas, in 1667, threw tobacco into the water offshore, so that the nimble little people living under the water would continue to furnish them with abundant material for their arrows (Donohoe, 1895, p. 103).

The medicine men of British Guiana teach that five classes of spirits direct the natural world and that each is represented by different stones (quartz, jadeite, etc.). Each neophytic medicine man is given two stones, representing the spirit and the spirit's wife, for his rattle. Provided the spirit is satisfied to be associated with the *piaiyen*, the stones breed and multiply, increasing the medicine man's powers in proportion as he has more spirits to assign to tasks or errands (Gillin, 1936, pp. 158–159).

THE INDIAN AS PROSPECTOR

The search for game and the need of roots, berries, and fruit as food made the Indian observant as to nature and kept him out of doors much of the time. Whether on a hunt or on the warpath, he was on the lookout for promising material for his arrows and his ornaments. Certain tribes, it will be remembered, were nomads and in a year's time covered a large area.

The Toltecs are reputed to have been great gem fanciers and Aztec tradition ascribes to the former their own knowledge of the art of working precious stones. The Toltecs attributed its invention to the god Quetzalcoatl (Biart, 1887, p. 325). The Toltecs were also reputed to be remarkable prospectors, for Friar Bernardino de Sahagún (1829–30, vol. 3, pp. 106–114) states:

Their knowledge of stones was so great, that, even though these were hidden deep in the earth, they discovered them through their natural ingenuity and knowledge, and they knew where to find them. Their manner of making such discoveries was the following: They would get up very early in the morning and go up to an eminence and turn their heads toward the place where the sun had to rise; when it rose, they carefully looked in every direction to see in what place any precious stone might be hidden; they would especially look for them in places that were damp or wet, and particularly at the moment when it was rising: then a slight smoke would go up quite high, and there they found the precious stones under the earth or inside of another stone, whence the smoke would issue.

In instances, at least, the Indian recognized the value of mineral indicators, for on the Coppermine River, northern Canada, the local Indians dug for copper in the detrital deposits "wherever they observe the prehnite lying on the soil, experience having taught them that the largest pieces of copper are found associated with it" (Richardson, *in* Franklin, 1823, pp. 528–530).

The efficiency of the Indian as a prospector is conclusively shown by plates 4 and 5, on which are indicated, though admittedly incompletely, the sources of the Indian's supply of gems and decorative stones. Note how many of the soapstone localities of the eastern States were known to him and how well the mica belt of the Southern Appalachians is delimited. Petrographers searching for glassy phases of igneous rocks might well be guided by the Indian's obsidian mines while practically all of the known turquoise localities of our own Southwest and the emerald deposits of Colombia were worked by Indians in prewhite time. As to nephrite and jadeite, the aborigines had sources which are still unknown to their white successors.

Many mineral occurrences were shown the whites by Indians, and Indian or half-breed prospectors have played no mean role in discovering mineral deposits in the United States and Canada in the past five decades.

MINING METHODS [1]

The Indian's metal mining technique and his knowledge of smelting were inferior to those of the Greeks and Romans and probably those of the Carthaginians. As a gem miner, however, he was about as skillful as the Egyptian and Asiatic peoples, the great gem miners of the Eastern Hemisphere in classical times. To us his knowledge necessarily seems crude, but in fairness to the Indian it should be compared with the European knowledge of the early sixteenth century, a comparison considerably less disadvantageous to the Indian.

We are dealing with a mining industry in its infancy, and in consequence the major portion of the Indian's mineral products came from river gravel, although marine beaches, glacial moraines, and weathered outcrops furnished a second part and hard rock mining still a third. Predominantly, the Indian was a placer miner. George Catlin told G. E. Sellers (1886, p. 874) that the Indians considered chalcedony, jasper, and agate most easily chipped into arrowheads and the principal sources of their supply were "the coarse gravel bars of the rivers where large pebbles are found." It is not unusual for one side of an artifact to be a segment of a pebble; for example, the jadeite of the Aztecs and Mayas.

Alluvial mining for gems and ornamental stones presumably consisted largely of visual inspection and hand sorting. The rudiments of gravity concentration in gold placer mining appears to have been known to certain of the Indians of both North and South America (pl. 1) but it could rarely have had application in procuring gems and decorative stones. On one of the Aleutian Islands, however, amber occurs in a steep bank of friable material. "The natives spread a walrus-skin between two boats at the foot of the bank and dislodge the earth, which falls upon the skin and from this debris much amber is obtained" (Dall, 1870, p. 476).

It is by no means impossible that silicosis existed among the Indians, for the Yana arrowhead makers of northern California dreaded to breathe obsidian chips. They believed these caused many diseases and it was a function of the medicine man to "suck" out such fragments (Pope, 1918, p. 117).

Mineral deposits, being rare and unusual phenomena essential to the natives' well-being, appealed to the religious mysticism of the savage. Spirits guarded them and because of the savages' animalistic religious viewpoint, these spirits were often fearsome and might be

[1] Holmes, W. H. Handbook of aboriginal American antiquities. Pt. 1. Introductory. The lithic industries. Bur. Amer. Ethnol. Bull. 60. 1919.

assisted by birds of prey or hideous snakes. On the Red River near Lake Winnipeg, Verendrye, writing in 1729, tells us (1927, p. 46) there was "a small mountain, the stones of which sparkle night and day. The savages call it the 'Dwelling of the Spirit': no one ventures to go near it." Similarly, the salt springs of Syracuse, N. Y., through fear of their spirit, were not used by the Indian (for a western instance, see Irving, 1888, p. 74). Again, as the deposits were the property of the spirits, compensation must be made for minerals extracted,— a votive offering, perhaps of a little tobacco or a bracelet. The Indian was unwilling to discover his mines to the whites lest the spirits punish him and the Navaho Indians even invented terrifying tales to keep the whites from their garnet mines. Pedro Pizarro (1921, vol. 2, pp. 393–394) tells us that the "wizards" tried to keep the Indians from showing Lucas Martinez the mine of the Sun, an admonition successfully backed up by an eclipse of the sun followed by an earthquake. Wabishkeepenas, who in 1820 attempted to show Governor Cass and H. R. Schoolcraft the large mass of native copper near the mouth of the Ontanagon River, was kept by the spirits from finding it. So incensed were the tribesmen at him that he was cast out of the tribe and almost starved to death (McKenney and Hall, 1933, vol. 1, p. 349). We may surmise, however, that some Indians sensed that with the discovery of mines, whites would appear, with consequent unsettlement of the Indian system of economics.

Due to such beliefs, certain of the Indian miners, at least, and perhaps all, performed religious ceremonies to propitiate the spirits or gods before beginning mining. This was true of the aboriginal emerald miners of Colombia, of the Plains Indians in mining catlinite, of the California Indians in quarrying magnesite, and of the Eskimo in mining soapstone and jade. Sugar Hill, a California obsidian locality, was sacred ground, whose spirit the Pit River Indians feared to offend. The Oregon obsidian arrow makers abstained from water while making arrows; if a blade broke, in the making, the spirits were against the maker and the broken pieces were thrown away never to be touched again; if a maker showed anger, his punishment would be twice as severe; in consequence, he sang a hymn of praise to the spirits when he started to make another point (Forbes, 1935, pp. 307–309). Peter Martyr tells us that among the Indians of Veragua, Panama, gold was sacred and mining was preceded by fasting and penance. The Indians of Colombia only mined what placer gold they needed and any surplus was returned to the stream; if they took more than their need the "river-god would not lend them any more" (Sumner, 1907, p. 142). The Pueblos while gathering salt at Salt Lake, near Estancia, N. Mex., were required to be quiet, silent, and serious. "If they speak or laugh or make fun they will stand just where they are and die" (Benedict, 1931, p. 7). The Hopi, for 4 days prior to starting out for the Colorado

River to obtain rock salt, observed a taboo on sex relationship. The Pueblos of Isleta (Parsons, 1932, pp. 320–321) get their red paint from the Manzano Mountains to the east of their pueblo. Proper religious ceremonies soften the hard rock so that it can be scraped out with one's fingers. One must not take more than he needs or it will return to its pristine hardness. In the West Indies religious ceremonies preceded placer mining and the miner for 20 days before starting on his expedition observed strict continence or else he got no gold (Joyce, 1916, pp. 67, 243–245). The Caribs of British Guiana get their best clay for pottery from a small hill near the mouth of the Cuyuni and long journeys are made to it. All mining is confined to the first night of the full moon and by break of day the natives are on their way homeward with a big supply. Pots made from clay obtained at any other time break and transmit disease to those who use them (Schomburgk, 1922, vol. 1, p. 203).

Many of the aborigines had, however, passed beyond the first stages of mineral exploitation, that of hand sorting the valuable component from stream gravels, morainal material, and outcrops, and actually attacked the ores and precious stones in place either by open cuts or by underground workings. The Indian, indeed, did much quarrying, the open cuts, particularly of turquoise and obsidian, being extensive.

Before searching for gold in vein outcrops, the Panama Indians first burnt the grass, thus laying bare the rock.

The Indian hard-rock miner used in primitive form most of the elements of modern mining. He had as mining tools stone hammers and sledges, deer- or caribou-horn picks and wood (one end often being hardened by being charred), horn, stone, and copper gads. Knowledge of breaking rock by building a fire against it and then throwing water upon it appears to have been widely spread (Lake Superior, southwest Wisconsin, North Carolina, and other southern mica mines, Pennsylvania jasper mines, Arkansas novaculite mines, the New Mexican and adjacent turquoise mines (pl. 2), Mexico lode mines, etc.). This great invention was a logical one as many Indians must have noted that the boulders, upon which they built their fires were fractured when, for safety, the embers were quenched with water. Most of the workings are pits or open cuts with their greatest dimension following the strike of the deposit. From them certain gem-bearing beds were stoped (Tylor, 1861). From such open cuts also, in instances, winzes extended, there being aboriginal workings in the Los Cerrillos turquoise mine to points 100 feet below the present surface. Short tunnels occur in the North Carolina mica mines and in the Mineral Park, Ariz., and San Bernardino County, Calif., turquoise mines. In the aboriginal salt mine, 3 miles south of St. Thomas, Nev., certain galleries are reported to be 300 feet or more long (Harrington, 1930). In the placer mines of the Chuchiabo district, province of Callao, Peru, Pedro Sancho,

Pizarro's secretary, states that some of the galleries were 40 brazas (about 240 feet) long (Means, 1917, pp. 163–165). In the ancient salt mine at Camp Verde, Ariz. (Morris, 1929, pp. 81–97) 4 or 5 different levels exist, each following a highly saline horizon. Two of these, 8 to 12 feet apart, are connected by a winze.

To carry out the product, hide and birchbark bags (Lake Superior copper mines and paint mines of Havasupai, Ariz.) were used and, in the Lake Superior open cuts, paddlelike wooden shovels for mucking have been found. In those mines, drainage was by ditching, supplemented by cedar troughs, and wooden bowls were apparently used to scoop out the water. The Havasupai of the Grand Canyon region, Arizona, obtain their red paint from a mine on Diamond Creek and to reach the portal on a cliff face they use ladders. A "chicken" ladder, a tree with its branches lopped off, was found in the old Lake Superior copper mines and another in Mammoth Cave. In the former district, too, rock pillars were left at one point and at a second the hanging wall was supported by huge granite boulders. There, also, great masses of native copper were raised by the use of levers and wooden props.

Certain features of lead mining by the Sauk and Foxes of the southwest Wisconsin lead district over 100 years ago are of interest although Indian methods may have been improved through intercourse with the whites (Meeker, 1872, pp. 271–296; History of Jo Daviess County, Ill., 1878, p. 836). There were then some 500 miners, largely women and old men for here, as opposed to most Indian mining, the able-bodied men were loath to mine. Some of the pits were 45 feet deep, the bottom being reached by a ramp. Ore and waste were removed by a mocock (a basket of birchbark or buckskin) which was dragged out by a rawhide rope. Rock was broken by the fire method. Drifts were run some distance into the side hills. The mining tools were originally buckhorns, but later European iron tools were introduced.

Normally, as little waste rock as possible was broken but, to get at plates of native copper in the Lake Superior region, the gangue on either side was, in places, removed and, in other open cuts, flint nodules were mined by undercutting. In turquoise mining in the Southwest, as the gem occurs as thin veinlets or small nodules, the ore was removed in blocks which were then carefully broken into small pieces to extract the gem. The ratio of waste to product was large. In mining gypsum in Mammoth Cave a circular area was cleared on the cave floor and the valuable gypsum sorted from the limestone waste (Pond, 1937, p. 178).

Most of the pits were sufficiently lighted by daylight but in Wyandotte Cave, Ind., mining was carried on over a mile from the cavern entrance. Flaming torches lighted the miners' work, as they did in the Camp Verde, Ariz., salt mine and the selenite deposit at Gypsum Cave, Nev.

On the Huallaga River, in eastern Peru, rock salt occurs in beds at the river's edge and the Indians make long canoe trips to obtain it. The overburden is first stripped and then trickles of water are led over the salt, which gradually dissolve their way to bed rock and the blocks are broken up into fragments of convenient size to carry away (Dyott, 1923, p. 130; Kerbey, 1906, p. 185).

That aboriginal mining had its major disasters is proven by the Pueblos' tales of miners, who were robbing pillars, being entombed in the New Mexican turquoise mines. The disaster is always brought on by an irreligious act either, according to variants of the tale, because of mining proscribed pillars or because a miner gave turquoise to his sweetheart, against which gift there was a taboo (Benedict, 1931, pp. 196–197, 236, 254). Alonzo W. Pond (1937) tells a dramatic story of an Indian gypsum miner trapped in Mammoth Cave. At Chuquicamata, Chile, in the pre-Hispanic pits, the mummified body of a woman was found, her head crushed by a fall of rock.

From the modern point of view, most of the open pits were small but, at Flint Ridge, Ohio; Magnet Cove, Ark.; Los Cerrillos, N. Mex., and Hidalgo, Mexico, the material removed must be measured in hundreds of thousands of tons. In the quartzite quarry, 125 miles north of Cheyenne, Wyo., Wilbur C. Knight (1898) estimates the tonnage of rock moved "by hundreds of thousands if not by millions of tons." The great majority of the mines listed (p. 56) were small-tonnage operations and most of them, at least, supplied the needs of a small number of primitive people. But also in Europe and Asia four centuries ago most of the mines were small operations.

"High grading" was feared at least in the Peruvian communal mining, for at the Chuchiabo gold placers, Pedro Sancho (Pizarro, 1917, 165) states "they [the caciques—S. H. B.] have taken such precautions in the matter that in nowise can any of what is taken out be stolen, because they have placed guards around the mines so that none of those who take out the gold can get away without being seen."

THE INDIAN'S KNOWLEDGE OF COMMERCIAL CHEMISTRY

The Indian knew something of chemistry; he burned limonite to produce red ochre for paint, gypsum was dehydrated to produce whitening, salt was obtained by evaporating sea water or saline spring water, and the Peruvian Indians smelted simple ores. The Kamia leached the salt-impregnated earth of Salton Sink and crystallized the salt out by boiling (Gifford, 1931, p. 4); this was done also by Indians of eastern Peru (Smyth and Lowe, 1836, p. 145), and by the Chibchas (Thompson, 1936, p. 120). A Potawatomi chief stated that their tribe had first noted elks drinking at salt licks: the Indians then tasted the water and, liking the flavor, boiled their meat and vegetables in it; finding this palatable "they boiled down the water in the manner that they had done the sap [i. e., hard maple sap—S. H. B.] and thus obtained salt" (Keating, 1824, p. 118).

The Tapuyas, of Brazil, made saltpeter by leaching earth containing it and then boiling the solution until the salt crystallized out (Warden, 1832, vol. 5, p. 209). Alunogen was used by the Navahos as a mordant.

THE EFFECT OF INDIAN MINING ON THE COMMERCIAL CONQUEST OF AMERICA

The hoarded mineral wealth of the Indians and the mines from which it had been obtained hastened to a remarkable degree the development of mining in America. Many of the earliest Spanish metal mines in Mexico and Peru were but further development of aboriginal mines. The natives' knowledge of the occurrence of gold, silver, emerald, and turquoise expedited mining development by many decades. Further, the primitive system of roads, the Indian trails, were followed by the white man in his exploration and conquest of the country.

INDIAN MINING LAWS

Usually the mine belonged loosely to the tribe in whose territory it occurred but in most cases working parties from other tribes could take what was necessary for their own needs. The mineral mined north of the Rio Grande, at least, was the personal property of the miner and he could use it or trade it as he saw fit (Gilmore, 1929, pp. 99–100; Weyer, 1932, pp. 174–176). In some instances, a valuable deposit for a time was sacred ground, open to all comers; for example, the Minnesota catlinite deposits (see p. 48), the Wisconsin catlinite deposits, and the blue clay of Blue Earth River, Minn., and probably the Obsidian Cliff obsidian (Alter, 1925, p. 381). Among the Pomo Indians of north-central California, the magnesite deposits and the obsidian quarries were operated, after proper votive offerings, by all the Indians of the region. Should hostile villages meet by chance at such places, their quarrels were forgotten; naturally, however, each had its own encampment and each party went about his business separately (Loeb, 1926, p. 194). Flint Ridge, Ohio, also is stated to have been neutral ground (Wilcox, 1934, p. 174). Neutral ground was not confined to mining for it is said that among the Araucanian Indians of Chile warfare ceased during the piñon nut season (Latcham, Jr., 1909, p. 341).

Such common use of mineral resources among the Indians probably originated through fear of angering the spirits of the mines. In certain instances, however, tribal rivalry existed as to the ownership of mineral deposits, and the Modak and Pit River Indians fought for the possession of the rich obsidian deposits west of Glass Mountain, Calif. Even monopolies existed in rare instances; for example, an old Natchez Indian alone made black marble pipes, nor did he permit his fellow tribesman to know the source of his raw product. For a common pipe he demanded "half the price of a blanket" (Schoolcraft, 1851–57, vol. 5, p. 692).

Gold, silver, and precious stones found in the Incan Empire were delivered as tribute to the Inca and he and his family wore them and also those nobles and captains he delighted to honor; they were used also in adorning the temples. "They were merely valued for their beauty and splendor" and were only mined when the Indians had no other work to do as "these were not things necessary for their existence." "Yet as they [the Indians] saw that these metals were used to adorn the palaces and temples (places which they valued so highly) they employed their spare time in seeking for gold, silver, and precious stones to present to the Ynca and to the Sun who were their Gods." (Garcilaso, 1871, vol. 2, pp. 21–22.) The Incan government, apparently with parental care, did not permit mining to be so extensively carried on by any individual as to injure his health.

TRADE

Due to less perfected methods of transport than our own, the stones used by the Indians were more likely to be of local origin than they are with us. In consequence, the source of precious stones was likely to be near its user's home—an aid in tracing its provenience but one to be used cautiously. Each tribe used the best stone his bailiwick afforded for the purpose required. A suitable mineral was much used by the tribe living where it occurred and from such centers gradually became less common until it disappeared where the limit of the local barter was reached. Conversely, if a region supplied no superlatively good material, for arrowheads, for example, the stones used might be of diverse origins. As the Indian liked variety and particularly brightly colored stones he was willing to barter articles of value for such as attracted his fancy. Much material from a distance was, however, used and, indeed, the Indians of the Argentine coast were wholly dependent on imports. In the first place, some of the tribes were nomadic and in tribal wanderings side trips were doubtless made to localities yielding desirable minerals. It is known that in 1680, a war party of Iroquois braves attacked tribes west of the Mississippi, distant from their New York home some 1,000 miles, and that other Iroquois war parties attacked the natives of South Carolina and of Lake Superior (Morgan, 1901, vol. 1, pp. 12–13). In some instances periodical trips were made for the particular purpose of procuring the desired material; we may cite the excursions to the pipestone quarry in Minnesota and the long wanderings of Eskimo after soapstone and other mineral substances. Roderick Macfarlane says the Eskimo "in singing and shouting boatloads" journeyed 400 miles to get flint from the quarries at Fort Good Hope (Stefánsson, 1922, p. 12; 1914, pp. 17–18). Again, by barter from tribe to tribe, some mineral substances almost traversed the continent until they found owners who treasured them too dearly to part with them. In the eighteenth century, Indians of the Northwest, even as far north as Montana, by intertribal barter possessed Spanish goods from New Mexico. Minnesota catlinite was carried as far as New York and Georgia; the Mound Builders had obsidian in quantity, probably obtaining it from the Yellowstone National Park (1,500 miles), mica and soapstone from the Appalachian Mountains (250 miles), and copper from Lake Superior (600 miles). New Mexican turquoise reached Mexico City and the Mayan cities, and Colombian emerald was so common in Peru that for at least two centuries after the Conquest it was known as Peruvian emerald. Certain tribes, for example, the Nez Percés, were outstanding as traders and over 100 years ago the Chippewas told William Cameron that they sometimes went as far as Virginia to barter Lake Superior copper for the products of the Atlantic coast (Fowke, 1888–89, pp. 402–403).

The Aztecs in particular, and the Mayas and Caribs to perhaps an almost equal extent, had a merchant class who journeyed far beyond the limits of their own countries. Colonel Church believes the Caribs traded along the seacoast of northwestern South America and the West Indies and probably in the entire Gulf of Mexico, including Florida (Church, 1912, p. 46). But the most interesting example of trade was the possession by the pre-Colombian Caribs, of the tiny island of Montserrat in the West Indies, of ornaments of amethyst, carnelian, jadeite, turquoise, rock crystal, chalcedony, lapis lazuli, nephrite, and cannel coal (Hodge, 1922, pp. 65, 75; Harrington, 1924, pp. 184–189). These semiprecious stones are all foreign to the island and strongly suggest that for their raw material the local Carib lapidaries were able to draw upon a number of different South American localities, certain of which must have been 2,000 miles away. Charlevoix, indeed, states that the Haitians have a legend that the green stones with which they hollowed out their canoes came from off the island and he specifies from the upper Amazon (Schomburgk, 1922, vol. 1, p. 264). C. F. P. von Martius also reports that West Indian Caribs spoke of their green stone amulets as "polished from the far-off continent" (Von Martius, 1867, vol. 1, pp. 731–732). [My attention was called to this reference by Miss Gladys C. Randolph.] The "trade trail" of nephrite along the Lesser Antilles to Cuba certainly suggests the South American origin of that stone. The materials reached the skillful Montserrat Island lapidaries unworked. Since many rocks and some minerals possess characteristics which indicate their source, it is suggested that petrographic examination of artifacts might throw much light on early American trade routes.

C. C. Jones (1859, p. 19) says that traditions then existed that arrow makers from the Georgia mountains in olden times left for the sea to trade their wares with the coastal peoples. Their avocation was honorable, they took no part in war, and were permitted to go wherever they pleased, being received everywhere hospitably. When, in 1584, Capt. Arthur Barlow and Capt. Philip Amados traded with the North Carolina coastal natives, they found the chiefs had precedence in the bartering and if they were present the commoner sort did not trade. When corn was ripe the Sioux arrived at the Hidatsa villages to trade and from "the time they came in sight of the village, to the time they disappeared there was a truce. When they had passed beyond the bluffs, they might steal an unguarded pony or lift a scalp and were in turn liable to be attacked" (Matthews, 1877, p. 27). In South America also, traders were permitted to traverse the country of their enemies in part because they carried with them the latest news (Im Thurn, 1883, p. 271).

Cabeza de Vaca could never have made his marvelous transcontinental trip had he not received fair treatment due to the commodities which he gathered and exchanged en route.

The wealth and variety of precious stones in the hands of the Aztecs was due in part from A. D. 1406 onward, according to Sahagun (1880, p. 547), to the Aztec traders who covered not only their own country but also penetrated the country of the surrounding tribes, traveling in the beginning of the sixteenth century as far south as Guatemala. As they had no beasts of burden, they packed their wares on their own backs, and it can be safely assumed that the cargoes they brought back with such difficulty to Mexico City were considered very precious. Owing to their familiarity with foreign tongues, these traders also served as imperial spies and frequently as the entering wedge of conquest.

As to when trade originated, we have no data but it presumably began soon after the Indians had stone artifacts. We can state, however, that it was well developed both in North and South America not long after the time of Christ (Coplico-Zacatenco culture, Mexico, the early Mayan Empire, Late Basket Makers, etc.).

In Mexico City, the Spaniards were surprised to find in the market a quarter given over to the goldsmiths who sold goldware and gem-set jewelry equal to or surpassing the handicraft of their Spanish contemporaries. Markets or fairs seem to have been held in many of the villages on set days which were attended not only by the people of the adjoining territory but by traveling merchants from afar. Bernal Diaz was greatly astonished "to see the vast number of people, the profusion of merchandise exposed for sale, and the admirable police system and the order that everywhere existed." Special magistrates held court and settled disputes on the spot; official weight-and-measure inspectors were present and falsification was severely punished (Joyce, 1914, p. 129–30).

The Mayas, particularly after they moved into Yucatan, inhabited a country without precious stones. However, they kept in commercial contact with their old home, the mountainous part of Central America, from which they got opals, presumably jadeite and probably other gem stones, but they doubtless obtained the majority of their stones by bartering with the Nahua peoples of Mexico. Indeed, in discussing the various articles used by the Mayas as currency, Cogolludo (1688, lib. 4., ch. 5) includes "certain precious stones and disks of copper brought from New Spain which they exchanged for other things." Spinden and Mason (Mason, 1926, p. 439) are convinced that they had emeralds from Colombia, although the writer has never seen an emerald in Maya jewelry. The Maya merchants, like the Aztec, traveled far and wide. Their gods were Xamen Ek, god of the North Star, and Ek-chuah, god of commerce, and to the latter when on the road they prayed nightly for safe return home. Much of their trade, like that of the Caribs, was doubtless by water as their canoes, manned by from 25 to 30 paddlers, made relatively long voyages. Columbus on his fourth voyage, in 1502, sighted such a trading canoe in the Caribbean off Bonacca Island.

For centuries, the Alaskan natives and those of northeastern Siberia have been in commercial contact. American soapstone, pectolite, and nephrite were traded with the Siberian natives (Kotzebue, 1821, vol. 3, p. 296). Alaskan nephrite is found in the ruins on St. Lawrence Island. On the other hand, Asian turquoise and amber (Weyer, Jr., 1929, p. 234) have been found in Aleutian graves.

GEMS MINED BY AMERICAN INDIANS

DIAMOND

Richard F. Burton (1869, vol. 2, p. 107) states that diamonds in Minas Geraes, Brazil, were "used it is said by the Indians as playthings for their children." While not susceptible of proof, the statement is not improbable because when the Portuguese first visited Brazil, the fact that the natives mined the associated alluvial gold was evidenced by their possession of gold fishhooks. From time to time an Indian hunter or miner must have been attracted by a diamond in a stream and picked it up, just as the Brazilian gold miners had done, prior to the recognition of the stones as diamonds in 1720. We may add that the fine diamond, "The Star of South Africa," was bought from an unsuspecting Negro sheepherder and that the Kashmir sapphires were used locally as gun flints—and they made good ones, too—before their true nature was known.   It is probable that the Indians knew of at least some of the precious stones of Minas Geraes before the white man arrived, for Master Antonie Knivet, who was with Thomas Cavendish on his second voyage in 1591, says that at a village inland from Rio, the Tamoyes found "stones as green as grasse, and great store of white glistering stones like Christall, but many of them were blew and greene, red and white, wonderfull faire to beholde" (Purchas, 1905–6, vol. 16, p. 220).   Still earlier, in 1572, due to information received from Indians, the Governor of Bahia sent Sebastian-Fernandez Tourinho on a long exploratory trip.   In the hinterland, he found different colored precious stones and the Indians told him of the existence of other varieties (Warden, 1832, vol. 5, pp. 27, 71).

Travelers in British Guiana repeat a story, doubtless a myth, that some of the native manioc graters have inset in them small diamonds rather than quartz crystals (MacCreagh, 1926, pp. 276–279).

CORUNDUM (RUBY AND SAPPHIRE)

J. H. Howard (1936, p. 78) states that "Mr. Burnham S. Colburn, of Bellmore Forest, N. C., has in his gem collection a ruby bead found in a Cherokee Indian grave in western North Carolina."   He apparently believes it to be of Indian workmanship and it may be noted that rubies occur in that part of the State.

At the Track Rock corundum mine, Union County, Ga., so named from an Indian pictograph nearby, blue and red detrital corundum is common.   As the Indians spent considerable time in the vicinity, it is conceivable that they collected some of the brightly colored corundum (King, 1894, p. 133).

EMERALD

Emerald was used ornamentally by the Indians of Colombia, Venezuela, Ecuador, Bolivia, Peru, Brazil, and Panama (pl. 4). Among the Peruvian Indians under the Incas, the emerald (called *Umina*) was the king of gems, even the turquoise being "not so much esteemed by the Indians" (Garcilaso, 1688, pt. 1, p. 341).

The Colombian emerald mines had been worked by Indians an unknown but long time before the Spaniards conquered the country, the Chibchas working the Chivor-Somondoco mines, and the Muzos the Muzo and Coscuez mines.   (Pogue, 1917, pp. 910–34; Pamphlet of Columbia Emerald Syndicate, 1921; Bauer, 1904; Olden, 1912, pp. 193–209; Benzoni, 1857, pp. 109–12; etc.)   The Chibchas, being great traders, also distributed the emeralds mined by the less civilized Muzos (Veatch, 1917, p. 249).   They bartered part of the emeralds with the tribes on the Magdalena below Neiva for gold (Joyce, 1912, p. 23). The Chibchas held fairs at which emeralds were featured, particularly that of Turmequé held every third day (Bollaert, 1860, p. 20).

So relentlessly did the conquistadores plunder the peoples of Ecuador and Peru that the large exports to Europe temporarily broke down the price structure of the emerald market. The fleet in which Father Joseph de Acosta returned to Europe (A. D. 1587) carried over 200 pounds of emeralds (Acosta, 1880, p. 226).

According to Joyce (1912, p. 42) emerald mining was inaugurated by religious ceremonies and was done in the rainy season, probably to take advantage of abundant water. Only the local Chibchas were permitted to dig at Somondoco and if a Chibcha permitted an outsider to do so, the former would die within a moon. J. Eric Thompson (1936, p. 120) states that the miners indulged in herbs causing them to see visions in which pay lodes would be revealed. The earth, excavated with pointed stakes, was washed in ponds, fed by deep-dug water ditches. W. H. Holmes (1919, p. 135) adds that the Colombian Indians also worked the solid-vein matter.

At the time of the arrival of the Spanish, emeralds in quantity were in the hands of the natives of northwestern South America and a few in those of Panama. The large number of emeralds in the hands of the Indians in the sixteenth century and their occurrence in old Indian graves in Peru and Colombia, support the thesis of a considerable age for the mines, as does the size of the pre-Hispanic workings at Somondoco. The Spanish chroniclers state that the keepers of the quipus said that Chimo Capac, a great chief of the Chimu period (perhaps 100 B. C.—A. D. 600) received emeralds among other tribute (Means, 1931, p. 64). William Bollaert (1861, p. 84), reports that the ruler or Scyris of the Caras, who conquered Quito about A. D. 1000, wore a large emerald, the hereditary emblem of his sovereignty. Further, Montesinos, a priest resident in America from 1628–42 (Montesinos, 1920, p. 94), states that emeralds were among the spoils of the Inca, Sinchi Roca, when he entered Cuzco in triumph after defeating the Chancas Indians (about A. D. 1100). It is stated that after Huayna-Capac conquered the Scyris he added to the *borla*, the insignia of Incan royalty, the emerald of the Scyris (Myers, 1871, p. 229). The mines are at least a thousand years old.

Recently S. K. Lothrop in his excavations at Coclé, Panama (1937, p. 186) found three emeralds; one, a pear-shaped double cabochon, was 1½ inches long and about 1 inch thick (189 carats), while a second weighed 112 carats. This is most interesting proof that the natives of Panama knew of the existence of emerald to the south as recorded by Francisco Lopez de Gomara (1749): "Some say that

Balboa received an account of how that land of Peru contained gold and emeralds.''

Many archeologists believe that the Mayas and Aztecs possessed emeralds, but to the writer's knowledge no emerald artifact has ever been found in any of the Mayan or Mexican ruins. I feel that the "emeralds" mentioned by the followers of Cortes were fine jades. Now, however, that S. K. Lothrop has found emeralds at Coclé, in Panama, a city which had trade relations not only with South America but with the cities to the north, the possession of emeralds by these people becomes less unlikely. Molina (1571) gives as the Aztec name of emerald *quetzdlitztli* (the stone of the brilliantly colored bird, the quetzál, *Pharomachrus mocinno*), an identification accepted by Pogue (1915, p. 109) and others (Squier, 1870, p. 246). I am inclined to believe, however, from Molina's description that this may well have been but a finely colored translucent jade.

While the South America lapidaries in some cases polished the stones into cabochon forms, they normally, like the Roman lapidaries, merely pierced the natural crystals so that they could be strung for necklaces or pendants. Emeralds in some instances were mounted by the Ecuadorian goldsmiths in gold jewelry.

BERYL

Beryl was not infrequently used by the Aztecs and it is one of the minerals mentioned in the Aztec Book of Tribute (Peñafiel, 1890, p. 79) as coming from the present State of Veracruz. It seems reasonable that some local Mexican source was known. It was used also by the Panamanians, Peruvians, and Brazilian Indians. At the Charleston Exposition in 1901, a crystal of golden beryl from an Indian mound, near Tessentee Creek, Macon County, N. C., was exhibited. It was 1½ inches in diameter and 2¼ inches long. Dr. G. F. Kunz stresses the fact that the Littlefield mine nearby produces fine aquamarines (1907, p. 45). D. B. Sterrett (1907, p. 799) reports that two beryl crystals were obtained from an Indian squaw at Lewiston, Idaho.

TOURMALINE

When the Mesa Grande, San Diego County, Calif., tourmaline deposits were found in 1898, it was reported that the Indians had long known of the deposit, a rumor substantiated by the occurrence of colored tourmaline crystals in Indian graves in the vicinity (Kunz, 1905, p. 23). The rediscovery of several deposits in the district was due to the Indians (Kunz, 1905, pp. 55, 56).

TURQUOISE [4]

Green and blue stones were especially valued by the Pueblos. (See p. 3.) They highly prized turquoise as an ornament, as a votive

[4] For further data on turquoise, see Pogue, Joseph E. (1915).

offering, and as a fetish; they decorated, in some instances, the lintels of their doors with it and used it as a measure of wealth and a means of investment. It may be added that the Navaho, to assist him in gambling, must needs have a fine piece of turquoise since Noholipi, the Gambling God in the Navaho origin legend, owed his remarkable winnings to a turquoise lucky piece (James, 1903, p. 150). Fray Geronimo de Zárate Salmerón (in New Mexico, 1618–26) says "to them it [turquoise] is as diamonds and precious stones" (Ayers, 1916, p. 217). In more recent times a string of turquoise fragments sufficient for an earring might well be worth the price of a pony (Blake, 1858, pp. 227–232). While Prof. J. S. Newberry (1876, p. 41) states that it was "so highly prized that a fragment of fine quality no larger than the nail of one's little finger and one-eighth of an inch in thickness was regarded as worth a mule or a good horse." He states that the Indians "discriminated accurately between the different shades of color" and were "not to be deceived by any base imitation." The value of turquoise beads was judged by the delicacy and purity of their blue color (Eleventh Census, 1890, vol. 50, p. 186).

The Pueblo Indians worked turquoise mines in our own Southwest at a number of places long before the Spanish arrived. Turquoise does not occur in ruins previous to those of the late Basket Makers and hence we can date the beginning of turquoise mining in the Southwest to about the fifth century (Roberts, 1929, p. 142). These people also apparently inaugurated the fascinating mosaic work of the Pueblo Indians. That the industry was an important one long before the discovery of America is shown by the many thousands of turquoise beads and pendants (30,000 in one room and 5,889 beads in a single burial) found at Pueblo Bonito, Chaco Canyon, N. Mex., dating from about A. D. 900 to 1100 (Pepper, 1909; Judd, 1925). It has furnished more turquoise ornaments than any other American ruin. George H. Pepper (1905, pp. 183–197) describes a turquoise pendant from Pueblo Bonito 3.4 centimeters ($1\frac{11}{100}$ inches) long and 2 centimeters ($\frac{78}{100}$ inch) broad at top and 2.5 centimeters (1 inch) at the bottom. A single mosaic cylinder 6 inches long and 3 inches in diameter was set with 1,214 pieces of turquoise. One can well agree (Pepper, 1920, p. 377) that the nearby Los Cerrillos mines must have been diligently worked by these people 1,000 years ago. John F. Blandy reported that in one grave near Prescott, Ariz., half a peck of turquoises worth $2,000 was recovered (Kunz, 1896, p. 910). Artifacts, indeed, suggest that over a thousand years ago the Pueblo peoples had greater wealth in turquoise than now and this opinion is strengthened by some of the Indian legends (Cushing, 1901, p. 385). Along the Salinas River, Ariz., ancient ruins are searched for turquoise after rains by modern Indians (Bartlett, 1854, vol. 2, p. 247). Turquoise is also prominent in the myths of the people of the Southwest

(i. e., the Zuñi, Pima, and Hopi), including the Navaho creation myth, suggesting the length of time it has been known to them. The extent of the commerce in turquoise in the sixteenth century is further proof that the mines were relatively old.

The most famous of these old turquoise mines, that at Los Cerrillos, N. Mex., the Tewa Indian place-name for which is the equivalent of "the place where turquoise is dug" (Harrington, 1916, p. 492), was reworked by the Spaniards until 1686, when the workings caved. From the size of the Cerrillos open-cut and of its dumps (pl. 2), and the large trees thereon, some of the latter being considered to be 600 years old, a considerable age for these pits is indicated. George H. Pepper, judging from country rock attached to some of the Pueblo Bonito turquoise pendants, is satisfied that Los Cerrillos was the mine from which they were obtained. In 1540 the Indians at the head of the Gulf of California told Captain Fernando Alarchon that the Pueblos dug the turquoises "out of a rock of stone" (Alarchon in Hakluyt, 1904, vol. 9, p. 300). Los Cerrillos workings have been described by many geologists. (Blake, 1858, pp. 227–232; Newberry, 1876, p. 41; Silliman, 1881, pp. 67–71; Kunz, 1890, pp. 54–59; Johnson, 1903, pp. 493–499; Sterrett, 1911, pp. 1066–1067, 1071.) Rather extensive ancient workings occur all over Mt. Chalchihuitl and on Turquoise Hill, 4 miles distant. Modern mining on Mount Chalchihuitl, according to Silliman, broke into open stopes in which were many stone hammers, some to be held in the hand and others grooved for handles. One of 20-pounds weight had the scrub-oak handle still attached by a withe. He adds that fire was used in breaking the rock. Sterrett says the "main pit on the northwest side of the hill" is "about 130 feet deep on the upper side and about 35 feet deep on the lower side, the rim is about 200 feet across and the bottom nearly 100 feet across. The large dumps of waste rock removed from this are about 150 yards long by 75 yards wide and from 1 to 30 feet deep. These dimensions do not correspond closely with those given by the earlier writers since this would give the dump an area of less than 2½ acres as compared with some 20 reported by Silliman." It does, however, suggest the removal of some 100,000 tons of rock. Silliman mentions aboriginal open chambers in solid rock, 40 feet long, and he states that at places modern mining has encountered aboriginal workings to depths of at least 100 feet. W. P. Blake says that 75 years ago it was visited by Indians from a distance and the Indians continue to this day to get turquoise at this mine. Although this mine is undoubtedly old, the Zuñi tale of the migration of the Turquoise Man and the Salt Woman suggests that, before the discovery of Los Cerrillos, turquoise was obtained from some mine farther north (Bunzel, 1932, pp. 1032–1033), possibly from the locality of La Jara, Colo.

There are also ancient Indian turquoise mines (Pogue, 1915, p. 55; Snow, 1891, pp. 511–512; Lindgren, Graton, and Gordon, 1910, p. 324; Zalinski, 1907, pp. 464–465, 474) presumably of pre-Columbian age in the Burro Mountains at several localities, on Hachita Mountains, and Jarilla Mountains (Hidden, 1893, pp. 400–402), Sierra County, near Paschal, N. Mex.; Sugar Loaf Peak, Lincoln County (anon., 1897, vol. 64, p. 456) and Crescent, Clark County, Nev. (Lincoln, 1923, p. 19; Vanderberg, 1937, pp. 24–25); Turquoise Mountain, Cochise County; and Mineral Park, Mohave County, Ariz. (pits and 20-foot long tunnels) (Blake, 1883, pp. 199–200; Mining and Science Press, 1902, vol. 85, No. 11, p. 102; Crawford and Johnson, 1937, pp. 511–522); Manvel, and Silver Lake, San Bernardino County, Calif., and near La Jara, Conejos County, Colo. (Kunz, 1902, p. 760). Several of the Burro Mountain localities known to white miners were discovered by evidence of aboriginal workings. The presence of charcoal indicates the use of fire in breaking the rock, followed by hammer work. The larger rock fragments were broken into small pieces in the search of turquoise. The stone hammers are "of rounded form, 4 to 8 inches or more in diameter and were evidently used without a handle" (Zalinski, 1907, vol. 2, pp. 464–465). Pits were sunk to a depth of 20 feet at least. Fire was also used at Turquoise Mountain, Cochise County, Ariz. Prehistoric turquoise mines occur over a large area in northwestern San Bernardino County, Calif., Manvel being one of the chief centers (Kunz, 1905, pp. 12, 107–109, 152–153; 1899, p. 580). Stone hammers and crude pottery occur in the old pits while pictographs are common in the vicinity. The pits, which occur in an area 30 or 40 miles in diameter, are from 15 to 30 feet across and up to 18 feet deep. Malcolm J. Rogers (1929, pp. 1–13) states that in the Silver Lake district, San Bernardino County, Calif., the pits occur in an east and west line 8 miles long and that short drifts were driven from the main pits.

The aboriginal trade in the turquoise of the Southwest (pl. 4) was widespread, extending from the West Indies and Yucatan on the south to Ontario on the north (New York Times, September 20, 1895) and from California on the west to Mississippi and Arkansas on the east. About 1527, Alvar Nuñez Cabeza de Vaca obtained turquoise from his Indian patients on the Rio Grande, Tex. The men of Hernando de Soto (A. D. 1542), when they arrived in the province of Guasco (eastern Arkansas), found in the possession of the Indians "Turkie stones—which the Indians signified by signes that they had from the West" (Purchas, 1905–7, vol. 18, p. 46). In a grave in Coahomo County, Miss., 100 turquoise beads and a small turquoise pendant were found (Peabody, 1904, vol. 3, p. 51). As glass beads were also found in the grave, this commerce with the Pueblos of the Southwest continued after the white man reached America. Vicente de Saldivar Mendoza in 1599 met, near the Canadian River in the vicinity of the present Texas-New Mexican line, a band of Apaches who had been trading for turquoise ("some small green stones which they use") with the Picuries and Taos pueblos of New Mexico (Thomas, 1935, p. 7). Agustin de Vetancurt writing between the years 1630 and 1680 says the Apaches visited the Pueblo of Pecos to trade for *Chalchuites* (1870–71, vol. 3, p. 323). Among the Apaches, precious stones had a directional symbolism: white shell, north; jet, east; turquoise, south; and catlinite, west.

When Sieur Perrot (1911, vol. 1, pp. 363–364; LeRoy, 1753), arrived among the Fox and Sauk in 1683, he was informed that the latter had visited lands to the Southwest where they met Indians coming from that direction whence they had brought "stones, blue and green, resembling the turquoise which they wore fastened in their noses and ears." They had also seen mounted men resembling the French, that is, the Spaniards of New Mexico. Twenty-one years earlier Radisson claimed that the Crees of the Lake Superior region had beads of "a stone of Turquoise" as nose ornaments which they obtained by barter from the "nation of the buff and beefe" (Burpee, 1908, p. 216). Malcolm J. Rogers (1936) states that a piece of turquoise was reported from a Wisconsin Indian site.

Jean Ribault (1875, p. 177) claimed that the French got turquoises from the Florida Indians in 1562 but his mineralogy may have been faulty or the material may have been derived from Spanish wrecks.

That turquoise, *Xiuitl*, was known to the Aztecs in early time is proved by the name of their God of Fire and Water, Xiuhcoatl ("The Turquoise Snakes") (Verrill, 1929, p. 185). Turquoise was rather commonly used by them, and also at the time of the Spanish occupation by the Indians of Sinaloa, Sonora, and Chihuahua. The latter certainly got theirs from our own Southwest, largely by barter with the Pueblos, although perhaps in part in mining excursions which they, themselves, may have made. A Franciscan friar (probably Fray Juan de la Asuncion) in 1538 found turquoise among the Indians of Northern Mexico which they got as day laborers' pay from the Pueblos to the north (Bandelier, 1890, p. 86) and the next year Friar Marcos de Niza repeats the statement. In consequence, at that time New Mexican turquoise in quantity was in the possession of at least the northern Mexican Indians and as, in the ruins of Pueblo Bonito sherds of Toltec pottery have been found, trade between our Southwest and Mexico is indicated even 1,200 years ago (Coolidge, 1929, p. 216). While the gem does occur in Mexico at several localities, none of the deposits are important and no aboriginal workings are known, although the Aztec Book of Tribute and the statement of Sahagun (1880, p. 77) that it was found in the "mines" indicate without much doubt that the Aztecs had Mexican sources of turquoise. Turquoise is said by Lorenzana to have been worked at Tollan, the capital of the Toltecs, about 1325 (Blondel, 1876, p. 294). In short, the Aztecs and their predecessors, the Toltecs, probably got their turquoise largely from New Mexico and Arizona but in part from local sources now undiscovered or exhausted. It appears to have been available in smaller pieces to the Aztecs than to the Pueblos and particularly was used as thin plates in mosaic work. Its treatment suggests its high value and possibly its foreign source.

That the Mayas had at their command considerable quantities of turquoise is shown by the 3,000 pieces set in the mosaic plaque found by Earl H. Morris at Chichen Itza in 1928 (1931, pp. 210–221). The turquoise, doubtless, was largely of New Mexican origin.

Turquoise was a popular and relatively common gem among the Peruvian Indians. It was also used by the natives of Montserrat, West Indies; Bolivia; Colombia; Ecuador; Argentina; and Chile. Indeed A. Hrdlička, in 1910, found a turquoise bead at Miramir, Argentina, on the Atlantic Ocean, 270 miles south of Buenos Aires. The natives of Ecuador had some fair-sized pieces of turquoise, a partially worked bead found on the Island of La Plata being 2½ inches long and 1½ inches in diameter (Dorsey, 1901, p. 266). No aboriginal South American mines are as yet certainly known but the gem occurs at certain places in the Andean regions of Peru, Argentina, and Chile. A suggestion as to a source is obtained from the name, Copiapó, which owes its origin, according to the Indian tradition to the great quantity of turquoise found in its mountains (Molina, 1809, p. 64). Further turquoise has recently been identified in the Chuquicamata copper deposit in Chile and as this mine was extensively worked by the Indians and as turquoise artifacts are found in the immediate vicinity, this was doubtless a source.[5]

D. Jenness found a turquoise bead some years ago in an ancient Eskimo grave on one of the Diomede Isles in Bering Strait, which he judged to be of Chinese origin from etchings on it.

Garnet

Garnet was used by the Pueblos, although it appears to have been one of the later stones known, as it has not, to the writer's knowledge, been found in Cliff Dwellers or Basket Makers ruins. Fray Gerónimo de Zárate Salmerón, who served in New Mexico from 1618–26, in his account of his missionary work states that in the Pueblo of Picuris (Taos County) "are garnet mines," a garnet locality known to this day (Ayer, 1916, p. 217). Today the Navahos collect the gem from their reservation as do other Arizona Indians from several localities, obtaining the gem both from the desert sands, from ant hills, and the source rock (Kunz, 1890, p. 80).

---

[5] Personal communication, L. W. Kemp.

Of doubtful correctness is the reported occurrence of garnet in Mound Builders graves (Shepherd, 1890, p. 103).

The Comanche Indians collected garnet at Jaco Lake, Chihuahua, Mexico (Bauer, 1904, p. 360) and it was commonly used by the Aztecs and was known to the Peruvian Indians.

Don de Ulloa and Don George Juan, Captains of the Spanish Navy, who traveled in Peru in 1734 (Pinkerton, 1808–14, vol. 14, p. 550) state that "rubies" are found in a river near the village of Azogues in the vicinity of Quito. "Indians wash them recovering pieces as big as a lentil and sometimes larger. . . . But the inhabitants, content with this piddling work, do not trouble themselves to trace the origin of the mine."

OLIVINE

The Navahos collect from their reservation olivines which occur with garnets in the desert sands. That the pre-Columbian Pueblos possessed olivine, I cannot definitely state, although they probably did (Winship, 1896; James, 1920, p. xv). Olivine and bronzite are incidental constituents of meteoric iron found in Ohio Mound Builders mounds (Shepherd, 1890, p. 101).

LAPIS LAZULI

Lapis lazuli was possessed by the Indians of Montserrat Island, Peru, Bolivia, Ecuador, and Chile (pl. 4). One of the largest pieces of lapis lazuli known (24 x 12 x 8 inches) was found in a Peruvian grave (Farrington, 1903, p. 202). Lapis lazuli is at present mined in Chile. By the Peruvian Indians, it was called *huinzo*.

SODALITE

Sodalite was used by the ancient Peruvians and by the Indians of Bolivia, Ecuador, and Argentina. Lapis lazuli, or more probably sodalite, was common at Tiaguanaco, Bolivia, where pounds of worked chips may be picked up, some as large as one's hands (Uhle, 1903, p. 95). In 1928 Fr. Ahlfeld (Ahlfeld and Wegner, 1932, pp. 288–96; Ahlfeld and Mosebach, 1935, pp. 388–414) found old workings for sodalite on the northern part of Serrania of Palca on the East Cordillera of Cochabamba, Bolivia. The mineral occurs as veins in a dike of nepheline syenite cutting Devonian sandstone. The works consist of open cuts and tunnels, the largest of the latter being 300 feet long and 16 feet high. The Jesuits worked the deposits after the Indians abandoned them, the earliest workings being pre-Inca. Microscopically and chemically the sodalite resembles that of the Inca artifacts (Brendler, 1934, pp. 28–31).

Lazulite

Lazulite was used by the ancient Peruvians and Bolivians.

Opal

The fire opal of Mexico was known to the Aztecs. There is now in the Field Museum of Natural History, Chicago, a head of the Mexican Sun God carved in fire opal, probably the work of an Aztec lapidary. Since the sixteenth century, it has been in European collections, and was one of the gorgeous gems owned by Philip Henry Hope. *El Aguila Azteca* ("the Aztec eagle") is a lovely Mexican opal weighing cut, evidently by an Aztec lapidary, 32 carats. It is carved in the form of an eagle's head and is reported to have been found about 1863. It is said once to have belonged to Maximilian (anon., 1937, pp. 97, 99, 101). Snr. Garcia (1936, p. 559) reports the finding of worked opals in the mountains southeast of Lake Chapala near Jiquilpan, Michoacan. The accurate word picture of the opal in Sahagun's Nahuatl text descriptive of Aztec lapidary work (Seler's trans. *in* Saville, 1922, p. 33) is well known.

Wm. T. Brigham (1887, p. 256) states that the Quiches, a Mayan tribe, in pre-Spanish times, used noble opals from Honduras commonly in their jewelry. The Mayas also had hyalite.

W. Reiss and A. Stübel (1880–87, p. 29 and pl. 80) recovered relatively large pierced opal beads from the Ancon, Peru, ruin. The mineralogic determination was by Prof. H. Fischer, of Freiburg.

Common opal was known to the Indians of British Columbia, Washington, Oregon, Wyoming, South Dakota, California, Panama, Brazil, and Argentina and to the Pueblos, Mayas, and Aztecs. That used in Oregon or Washington in part is, or approaches, gem quality.

Distribution of Quartz Gems

Rock crystal and quartz, jasper and chalcedony were used by practically all tribes. Agate was known to the Indians of British Columbia, eastern Canada, the North Atlantic States, Virginia, the Mississippi Valley, the northern plains, the western mountains, the American Pacific coast, and the Pueblos. It was also used by the Aztecs, the Mayas, and the Peruvians and the Indians of Costa Rica, Panama, Colombia, Bolivia, British Guiana, Chile, Argentina, and Brazil. Carnelian artifacts have been found in Georgia, Illinois, the western mountain States, California, Oregon, and Washington, and among the Pueblo ruins. It was used by the Aztecs and the Mayas and the Indians of Costa Rica, Panama, Montserrat Island, Lesser Antilles, Colombia, Venezuela, Ecuador, British Guiana, and Brazil. In certain Colombian graves, as many as 8,000 beads, largely carnelian, have been found together with pebbles of carnelian, suggesting a stream origin. Such beads are readily sold to the present-day Indians living to the east, and a brightly colored one may be worth a mule. The local source is not sufficient for today's demand and beads are actually imported from Germany to satisfy it (Mason, 1936, pp. 212–216).

In North America amethyst was used by the Eskimo and by the Indians of eastern Canada, the southeastern States, the upper Mississippi Valley, and California, and by the Pueblos and Aztecs. It was also used by the aborigines of Costa Rica, Panama, Montserrat Island, the Lesser Antilles, and Peru. Smoky quartz was used by the Eskimo, the Indians of Newfoundland, Rhode Island, the southeastern States, upper Mississippi Valley, Colorado, Washington, Oregon, and California, and the Pueblos and Peruvians. Other species of quartz were used as follows: Moss agate (Saskatchewan, New York, southeastern States, Wyoming, Colorado, Texas, Utah, California, Oregon, and the Pueblos); rose quartz (New Jersey, Maryland, Virginia, Georgia, South Dakota, Aztecs, and Brazil); gold quartz (Georgia, California, and Arizona); citrine (Georgia, upper Mississippi Valley, and Dakota); prase (Pueblos and Aztecs); bloodstone (Oregon, Aztecs, Panama, and Peru); chrysoprase (Peru, Colombia, and perhaps California); iris (Mound Builders); Aventurine quartz (Aztecs and Mayas); plasma (Aztecs and Panama); and onyx (southeastern States and Washington).

AMETHYST

The Aztecs had some amethyst of fine color; finer, I am inclined to believe, than that of any of the Mexican sources we now know. Sahagun (1880, p. 771) mentions Aztec amethyst mines.

ROCK CRYSTAL

Quartz was doubtless obtained largely in river, marine, and glacial gravels but pits were sunk on quartz veins in New England, New York, and in the Appalachian Mountains of Virginia and North Carolina and it must have been a byproduct of mica mining. Quartz was, indeed, quarried at many places in the Piedmont region of the southeastern States, but since here, as elsewhere, it occurs so frequently as pebbles, gravels were the main source (Holmes, 1897). Master Alexander Whitaker, minister to Henrico Colony, Virginia, writing in 1613 (Purchas, 1905–7, vol. 19, p. 112) says that 12 miles above the falls on the James River is "a Christall Rocke wherewith the Indians doe head many of their arrows." When Albert de la Pierria founded Beaufort, Fla., about 1563, the Indians brought the Frenchmen presents of "pearls, crystals, silver, etc." In 1587, the Frenchmen of Charles-fort were given by the chief, Ovade, pearls, fine crystal, and silver ore said to come from 10 days' journey inland (Georgia probably) where "the inhabitants of the countrey did dig the same at the foote of certaine high mountaines where they found of it in very good quantitie. Being joyfull to understand such good newes and to have come to the knowledge of that which they most desired" the Frenchmen returned to their fort (Laudonniere, 1904, pp. 481–482). When John Verarzanus, a Florentine writing in 1524, states that the Florida natives were the possessors not only of crystal but also of turquoise, we become skeptical of his mineralogic attainments (Hakluyt, 1850, p. 106).

The Hot Springs, Ark., rock crystal locality was as well known to the Indians as to present-day mineralogists. Rock crystal and arrowheads made of it are common in Arkansas Indian graves. In Cavelier's account of La Salle's voyage, he says "about 50 leagues from the spot where we were [mouth of Rio Bravo, Tex.] were two or three mountains on the banks of a river from which were taken red stones as clear as crystal," possibly a distorted reference to Hot Springs (Shea, 1861, p. 28).

Father Gravier (1900, pp. 138–141) in his voyage on the Mississippi in 1700 "found in a small basket," in a temple in the village of the Taensas, a subtribe of the Natches Indians, "a small piece of rock Crystal."

The Navahos are stated to light their ceremonial fire from the sun by means of crystal (Curtis, 1907, p. 53). The Pueblo Indians of the upper Rio Grande during their rain ceremonies beat the drum to imitate thunder, and rubbed together pebbles of white quartz to produce an incandescent glow simulating lightning (Jeancon, 1923, p. 68). At Pecos, New Mex., A. V. Kidder (1932, p. 93) found a cylinder, set in a rectangle with a shallow groove into which the cylinder exactly fitted, both of white vein quartz. The cylinder is about 3 inches long and 1½ inches in diameter. Knowing that "lightning sets" still were used in religious ceremonies at San Ildefonsa, that night he rubbed the cylinder in the groove and finally the stones "became visible in a strange pale glow which flickered and died for all the world like distant lightning." Here we have a perfected machine perhaps 700 years old; the first Indian to observe the luminescence of quartz must have done so centuries earlier.

The Hopi used rock crystal in religious ceremonies, as well as in divinition (Fewkes, 1898, p. 730; 1904, pp. 107–109). It was also used in the religious ceremonies of the Zuñi, being, for example, placed on certain fraternal shrines, and by other Pueblos it was used to reflect the sun into kivas and into medicine bowls. Rock crystal was used in diagnosing disease both by the Pueblos (White, 1932, p. 110; Parsons, 1932, p. 285) and the Pima (Parsons, 1928, p. 459). Rock crystal is a common charm among the Yuma Indians (Forde, 1931, p. 196). The Australian medicine man paralleled his American confrere in many uses of rock crystal in the curative art (Goldenweiser, 1922, pp. 105–107).

According to a Shasta legend, a long time ago there was in the East a white and glistening firestone like the purest quartz. The coyote brought this to the Indians, and thus fire originated (Bancroft, 1886, p. 547). Quartz was supposed by the Chippewas to protect its owner against thunderstorms as the thunderbird would no more hurt it than a hen "the egg she has laid" (Densmore, 1929, p. 113). The most prized possession of the Cherokee medicine man was a rock-crystal-like mineral once embedded in the head of the Horned Serpent so prominent in Iroquois mythology. It was invaluable in treating the sick and foretold which of the braves should shun certain raids as their deaths were shown by it to be probable (Oldbrechts, 1930, pp. 547–552). Among the Ojibways white flint was called *mik kwum me wow beek*, or ice stone and as the name was also doubtless applied to rock crystal, it parallels the Greek from which our word crystal is derived. Similarly certain Alaskan Eskimo believe rock crystals are the centers of ice masses so solidly frozen that they become stone: they are, therefore, prized amulets (Nelson, 1899, p. 446). A Mandan medicine man, among other wonderful performances vouched for by white witnesses, could roll a snowball in his hand "so that it at length becomes hard, and is converted into a white stone, which when struck emits fire" (Maximilian, 1843; Thwaites, 1904–07, p. 342). Near the village of Lansingburgh, N. Y., is Diamond Rock, a mass of Quebec sandstone containing innumerable quartz crystals which glitter in the sunlight. According to Mohawk tradition, these are the joyful tears of a devoted mother upon her reunion with a wandering son (Sylvester, 1877, pp. 207–220).

Sahagun (1880, p. 771) reports that the Aztecs had crystal mines and Clavigero (1807, p. 16) states that these come from the mountains on the Gulf coast between Veracruz and Coatzacualco River, that is, those of Chinantla and the Province of Mixtecas.

The present-day medicine men among the Yucatan Indians pretend that they can see hidden things with the aid of rock crystal, and it is successfully used in the diagnosis of the ills of their patients (Mendez, 1921, pp. 173–174). When Hans Stade was about to be eaten by his eastern Brazilian captors, an old woman of the tribe shaved off his eyebrows with a rock-crystal razor (1874, p. 63). The Venezuelan Indian lover must shape for his beloved, as a betrothal gift, a cylindrical bead of rock crystal to be worn around her neck (Spence, 1878, vol. 1, p. 81). Spruce (1908) emphasizes the far-flung trade in rock crystal along the Amazon.

AGATE

Agate adorned the central mountain where, according to the Navaho creation myth, their tribe was created. The pebbles in the rattle of the Guiana medicine man are agate (Bancroft, 1769, p. 311).

JASPER

In Bucks, Lehigh, and Berks Counties, Pa., there are some nine groups of jasper quarries. A depth of at least 14 feet was attained. Fire was used in quarrying. A tree in the bottom of one pit shows the quarries must at least antedate 1680–90. At one place almost 40,000 cubic yards were excavated (Mercer, 1894, pp. 80–92; Deisher, 1932, pp. 334–341). Jasper was also quarried in Chester and Lancaster Counties, Pa. Flint Ridge, in Licking and Muskingum Counties, Ohio, is the site of aboriginal flint quarries, affording also jasper and chalcedony, which cover an area 2 miles square. In the cavities of the flint are quartz crystals up to 1 and 2 inches across. They vary from limpid to almost black. The pits are up to 80 feet in diameter. Fire and water were used to supplement the stone hammers (from 6 ounces to 20 pounds in weight) in breaking up the rock (Wilson, 1898, pp. 868–871; Fowke, 1888–89, pp. 517–520).

Sahagun (1880, p. 771) mentions Aztec jasper mines. Clavigero (1807, p. 16) states that jasper was quarried in the Mountains of Calpolalpan, east of Mexico City. R. H. Schomburgk (1846, pp. 28–29) states that jasper resembling verd antique is obtained by the Arecunas Indians from Mount Roraima on the Caroni River and is not only used but is traded to other tribes.

CHALCEDONY

Jasper and chalcedony occur as nodules in the quartzite of the Converse County, Wyo., quartzite quarry and were a byproduct of quarrying there (Dorsey, 1901, pp. 237–241).

CHRYSOPRASE

Near the California chrysoprase localities on Venice Hill, Tulare County, are depressions which D. B. Sterrett (1909, pp. 753–754) seems inclined to believe are aboriginal pits.

IRIS

Among the Mound Builders artifacts found at Mound City, Ross County, Ohio, were arrowheads of "transparent or hyaline quartz which from the brilliant play of colors upon their fractured surfaces are real gems" (Squier and Davis, 1847, p. 213).

AGATIZED WOOD

Agatized wood was used by the Indians from Oregon and Wyoming southward to Yucatan, Mexico.

The Petrified Forest, near Adamana, Ariz., was, to the Indians and particularly to the Pueblos, a source of agatized wood, amethyst, smoky quartz, and other members of the quartz family. Agatized wood was traded well out into the plains and southwest far into Mexico. Charles F. Lummis (1892, pp. 20–27; 1925, pp. 109–121) believes the Indians made special trips to gather chips from the shattered tree trunks. Agatized wood was used by the Basket Makers, but as neither they nor the Cliff Dwellers are known to have had either amethyst or smoky quartz, intensive exploitation of the Petrified Forest perhaps did not long antedate the coming of the Spaniards. Among the Arizona Indians, it was called *Shinarump*. Agatized wood, chalcedony, and other quartz minerals were also obtained by the Pueblo peoples from a "petrified forest" 25 miles south of Chambers, Ariz., on the Sante Fe Railroad (Roberts, 1931, p. 5). The agatized wood used by the Wyoming Indians was presumably of local origin, as sources are numerous in the West.

Artifacts of silicified or agatized wood are found also in Florida, Mississippi, Arkansas, Texas, and the upper Mississippi Valley. More or less beautifully silicified fossils were also rather widely used (silicified coral, New York, southeastern States, upper Mississippi Valley and Pueblos; shark's teeth, southeastern States; *Baculites*, Kansas; shells, Mound Builders, Pueblos, and Mayas; and crinoid stems, Pueblos). On account of their forms they were in high repute as charms.

DISTRIBUTION OF JADE

Jadeite was commonly used by the Aztecs and other Mexican tribes, the Mayas, all Indians of Central America, the West Indians and the Peruvians; it was also used less commonly by the Eskimo and British Columbia Indians, the Mound Builders, the Pueblos, and the peoples of British Guiana, Ecuador, and Brazil (pl. 4). Nephrite was commonly used and highly prized by the Eskimo and the Indians of British Columbia, Venezuela, Colombia, Ecuador, and Brazil; it was also known to the Indians of Oregon and Washington, the Pueblos, the Aztecs, Mayas, the ancient Costa Ricans, and the Indians of Montserrat Island, Cuba, and the Lesser Antilles, and of Argentina and Chile. The single occurrence known to me among the Peruvian Indians is reported by Uhle in the Ica Valley. It was also probably known to the Haitian Indians (pl. 4). Jade was early used by the American aborigines. A jadeite tablet found in Guatemala bears in Maya numerals the date equivalent to A. D. 60, while a Maya statuette from Veracruz has the date corresponding to 98 B. C., by the Spinden correlation. The latter has been usually considered nephrite but more recently was determined to be jadeite (Washington, 1922, pp. 2–12). The annals of the Cakchikels Indians of Guatemala, a Maya tribe, stated that jade was used as tribute in the second of the four Tulans from which the four clans came. The researches of F. W. Clarke and G. P. Merrill (1888, vol. 2, 115–130) lead them to believe that all the Alaska jade is nephrite while that of the ancient Mexicans and Mayas is largely jadeite although nephrite also occurs.

Nadaillac and Moorehead report jade in Mound Builders' mounds and jadeite is reported by Putman from a mound of these people in Michigan. (Putnam, 1886, pp. 62–63; W. Moorehead, 1917, p. 20; de Nadaillac, 1884, pp. 107–109.) This is most interesting, proving the extent of tribe by tribe barter, while suggesting, if we let our imagination run riot, a more direct commercial connection with the Mexican civilization or, via Florida, with that of the West Indies. Jadeite, while in the possession of the northern Mexican Indians (Chihuahua), was exceedingly rare among the Pueblos although A. V. Kidder and S. J. Guernsey (1919, p. 148) report a jadeite pendant in a Cliff House Ruin in northeastern Arizona. These authorities know of no other occurrence of jadeite in the Southwest, although it is also reported from Casas Grandes in Chihuahua. Nephrite also was very rare among the Pueblos although a small round tablet from Cochise County, Ariz., made of "an impure variety of nephrite" has been found (Holmes, 1906, p. 108). Warren K. Moorehead (1910, vol. 2, pl. 51) states that a jade effigy of a fish was found in a Pueblo ruin near Mesa, Ariz. Julian H. Steward (1937, p. 72) found a nephrite scraper in a cave on Promontory Point, Salt Lake, Utah. As much Southwestern turquoise reached Central Mexico and Toltec pottery occurs in the ruins of Pueblo Bonito, it would appear natural that jade would have been obtained from the south, particularly as the Pueblo peoples were passionately fond of green stones. The situation is perplexing. Did the Pueblos have a taboo against jade or was its export from Mexico northward prohibited by tribal decree? It reminds one of the rarity of amber among the Egyptians and early Mesopotamian peoples. Strangely enough, the Pueblos depended largely on local sources for their precious stones.

JADEITE

Jade (largely jadeite) was used early by the Mexican peoples. According to Toltec chronicles, Chimalman, the mother of the king Quetzalcoatl Chalchiuitl (about A. D. 839), became pregnant from swallowing a *chalchihuitl* (jade) (1886, vol. 5, p. 257). In the advice of a Toltec father to his son, the gods listened to the prayers of the wise men of old, "because they were of a pure heart, perfect and without blemish like Chalchihuitl" (Charnay, 1887, p. 179).

Jade (*chalchihuitl* in part; the term also covers turquoise and other green and blue stones) was highly prized by the Aztecs, and Father Sahagun says they could only be worn by the nobles or ruling classes. To show that the Aztecs valued it highly, I need but to remind you that when Montezuma and Cortes gambled, the native chieftain first paid his debts in gold but on the second night he promised the Spanish marauder something much more precious; this, to Cortes' disgust, proved to be four small carved jade beads. Montezuma had valued each one at two loads of gold, although I do not remember that the size of the load was specified. That it was relatively abundant among them is shown by the fact that Dr. H. M. Saville in 1900–1901 excavated the site of an ancient Aztec temple in Mexico City, and found therein over 2,000 jadeite objects.

It is not unusual for Aztec and Central American jadeite carvings to retain on the reverse side a segment of the original pebble from which it was cut. The jadeite was evidently of alluvial origin. Friar Bernardino de Sahagun says *chalchihuitl* was found in Mexico. Mrs. Zelia Nuttall, from a close study of the Aztec tribute rolls (1901, pp. 227–238) in which jadeite is listed as the tribute of certain towns, concluded that the material was obtained from a number of places in southeastern Mexico, the country east of a north to south line drawn through Mexico City, Chiapas, and Guerrero being particularly likely sources. Almost a generation thereafter her predictions were verified by discoveries of jade in Zimapan (Davis, 1931, p. 182) and as pebbles in rivers in Oaxaca and Guerrero (Caso, 1932, p. 509).

Jadeite was the most precious of Mayan possessions, and its ownership an insignia of wealth or power. A piece of jade was put in the mouth of the dead, curiously analogous to the Chinese custom. The Mayas had jadeite in relatively large pieces as some pectoral plaques of it are 6 inches square. In March 1937, the Carnegie Institution reported that Dr. A. V. Kidder had found in a pyramid mound near Guatemala City a water-worn boulder of jade 16 inches in diameter and weighing nearly 200 pounds. This beautiful light-green jade is certainly one of the largest ever found in America. A highly polished sphere of jade $1\frac{1}{16}$ inches in diameter, once used as a conjuring stone, was found at Chichen-Itza in 1928 (Morris, 1931, p. 210). Mayan jade, in part at least, was derived from stream gravels, for at Copan, Honduras, a human figure was engraved on one side of a pebble 8 inches long, the reverse distinctly showing its pebble form (Gann, 1926, p. 183). Guatemala and Costa Rica are likely sources in addition to trade with Mexico. H. J. Spinden (1913, p. 145) reports that some of the jades found at Monte Alban, Mexico, appear to be of Mayan workmanship, and were doubtless obtained by barter with the Mayan merchants. In later times, the Mayans were supplied with jadeite not only from Guatemala but also from Mexico. Thomas Gann (1925, p. 274) states that jadeite artifacts were more common in the early days of the Old Empire than in the later days of that era or in those of the New Empire, suggesting, perhaps, the partial exhaustion of a local alluvial source. The conclusion appears probable as in late Mayan times some of the artifacts are patently cut from larger ornaments of an earlier date.

Jadeite was widely used in Costa Rica, although it is much less common in the highland than in the Pacific coastal graves (Hartman, 1901, p. 171). At Las Guacas, Nicoyan Peninsula, Costa Rica, jadeite occurs more abundantly than anywhere else in the American continent, the finds being numbered by hundreds. Included in them are partially worked pebbles and unworked blocks. The Nicoyans cut jade cleverly and had a considerable trade in it both to the north and south. A stream source must be nearby (Hartman, 1907, p. 85).

NEPHRITE

Lt. J. C. Cantwell in 1884 heard of nephrite on the Kobuk River, Alaska, but the natives through superstitious fear of the mountain refused to guide him to the mine. Lt. George M. Stoney, however, in 1886 found the aboriginal nephrite mine at Jade Mountains on the north side of Kobuk River (1900, pp. 56–57). His Indians refused to visit the place for fear they might not return as "only the medicine man could visit it and then not until after a long fast" (Stoney, 1900, p. 56). Shungnak, the local Eskimo name for jade, is given to one of the tributaries of Kobuk River (Smith and Mertie, 1930, p. 345). In addition to jade in place, the pebbles of the rivers of the region were doubtless collected by the Eskimo. The natives also collected nephrite boulders occurring in the Fraser and Thompson Rivers, British Columbia, and on the high benches of the Fraser River (Camsell, 1912, p. 606), on the beaches of Puget Sound, and in southern Oregon. Nephrite boulders also occur on the upper Lewes River, Yukon Territory, not far from the eastern boundary of Alaska (Dawson, 1887; 1888, p. 186) and on the Rae River in the Coronation Gulf region (Jenness, 1925, p. 432) although we do not know that these sources were exploited by the Indians. Nephrite was used by all the Eskimo of Alaska, and they often made long journeys to procure it. By barter it had reached the Eskimo of the west coast of Hudson Bay and Baffin Island as early as the "Thule stage of culture" (say A. D. 600–1600) (Mathiassen, 1927; 1927 a, pt. 2, p. 27). Nephrite artifacts are found also in the Eskimo ruins of Newfoundland (Jenness, 1932). In consequence, the finding of an axe supposed to be of jade at Balsam Lake, Ontario, is not surprising (Laidlaw, 1897, p. 85; 1897 a, vol. 19, p. 69).

Indeed, jade was traded in all along the northern Pacific coast; the British Columbian sources furnishing material for the coast from the Straits of Juan de Fuca to Bering Bay and Jade Mountain, from the Aleutian Islands to the mouth of the Mackenzie River (Emmons, 1923). Of the various stone amulets, jade was the most valuable and among the Eskimo the stone had magical properies. A small bead was worth six or seven foxskins. A small adz blade among the Tlingits was worth two or three slaves and while a Tlingit was using it, his wife refrained from frivolity lest the blade break. The Tlingit name was *tsu* (green), a close approximation to the Chinese *Yu*. The British Columbian father handed a nephrite tool down to his son as a priceless heirloom. Among the Salish Indians it was called Stoklait (green stone).

The material of the jade (nephrite) ornaments in the possession of the British Guiana Indians is said by the natives of San Carlos to come from the source of the Orinoco River and by the Indians of the missions of the Caroni and of Angostura from the headwaters of the Rio Branco. The two localities are near one another (de Humboldt, 1814–29, vol. 2, pp. 395–402, 462). Tubular beads and labrets of this material were an object of barter through much of Brazil and British Guiana. Such are reported to have been used "as current money (Keymis, *in* Hakluyt, 1904, vol. 10, p. 491). Sir Walter Raleigh (1595) reported that the tribes of the Amazons traded jade ornaments for gold and the English exported them to England to cure kidney diseases as early as 1604 (Pinkerton, 1812, vol. 12, p. 283).

That the alluvial nephrite locality of Amargoza, Bahia, Brazil, was worked by the Indians seems probable from the abundance of nephrite artifacts in that region.

MINERALS AND ORNAMENTAL STONES MINED BY AMERICAN INDIANS

ACTINOLITE

Actinolite, or a rock largely composed of it, was a material for axes prized by the ancient Hopi and Zuñi (Hough, 1903, p. 322). It was also used by the California natives.

CHLOROMELANITE

Chloromelanite tools were particularly used by the Mayas of Guatemala, and by the ancient people of the Valley of Mexico and the State of Guerrero (Hodge, 1922, vol. 3, p. 47). It is presumably of local origin. It was also used by the pre-Spanish Colombians and Chileans.

PECTOLITE

From a pale greenish or bluish slightly translucent pectolite, the Eskimo of Point Barrow make hammerheads. Both pectolite and jade are called *Kaudlo* and are said to come "from the East, a long way off" (Murdoch, 1892, p. 60; Clarke and Chatard, 1884, p. 20). The Alaskan natives traded pectolite with the natives of Siberia. It was also used by the British Columbian Indians.

SERPENTINE .

Serpentine was used by practically all Indian tribes.

STAUROLITE

While the writer has not been able to definitely prove the report, it is probable that the Indians of Virginia used the local staurolite as an ornament. Such a striking mineral, and one so locally abundant could scarcely have escaped their notice.

MAGNESITE

The Pomo Indians obtained magnesite at White Buttes, near Cache Creek, and at Sulphur Bank, Clear Lake, Calif. This was made into beads and baked, in which process the color changes from white to an attractive buff or salmon color. As elsewhere stated, this served as money and was traded from the coast to the Sierra Nevadas (Kroeber, 1925, p. 249; Loeb, 1926, p. 178). The magnesite of the Pueblo Indians may also have been of California origin. It was also used by the Indians of British Columbia.

ALABASTER AND STALACTITIC CALCITE

Alabaster was used by the Indians of Canada, the North Atlantic States, Georgia, the Mississippi Valley, the Rocky Mountains, California, Cuba, Puerto Rico, Ecuador, Colombia, Bolivia, Brazil, Argentina, and Chile, and by the Pueblos, Aztecs, Mayas, and Peruvians.

The Indians found that Wyandotte Cave, Crawford County, Ind., contained two desirable products, a jaspery flint and stalactites of satin spar. They carried on mining a full mile within the cave, lighting their labor with flaming torches. From the lenses of flint protruding from the limestone walls they hacked flint flakes, with granite hammers, and also cut from a giant stalactite some 1,000 cubic feet of glistening alabaster. The imprints of their moccasins were still visible on the floor of the cave 80 years ago. They also dug down from the surface in one place until the cave formation was encountered and mined alabaster open cut. Deer antlers were used as picks in this work. (Fowke, 1922, pp. 108–109; Blatchley, 1897, pp. 156, 165–169; Mercer, 1895, pp. 396–400.)

The Aztecs quarried at Tecali, in Puebla, Mexico, what we now call Mexican onyx, and what they called *tecali*, probably derived from *teocali* or "lord's mansion" (anon., 1891, p. 729). It was widely used for images and vases and as windows in their temples (Fortier and Ficklen, 1907, p. 190).

GALENA

The brilliant luster of galena appealed to the Eskimo, the Indians of British Columbia, Canada, Florida, the upper Mississippi Valley, Virginia, Mississippi, Utah, the Yavapai Indians of Arizona, and the Pueblos. The Wisconsin-Iowa-Illinois lead district was known to the Indians before the white man's arrival and Nicolet (1634) and Radisson and Groseilliers (1658–59) were told of the deposits. Miami Indians brought to Perrot upon his arrival in the region in 1684 "lumps of lead" found, they said, in rock crevices (Neill, 1858, p. 139; Carver, 1778, pp. 47–48). While the writer agrees that smelting was taught the Indians by the French, he cannot agree with those who state that mining was taught them by the French as galena is widely and abundantly present in the prehistoric mounds of the upper Mississippi Valley. When, in 1810, two St. Louisianians, Colonel Smith and Mr. Moorhead, purchased the Wisconsin-Iowa-Illinois lead fields from Augustus Choutou, the Sauk and Foxes ran them off the property. Fearing the effect of this action on the American Government, the tribes at once sent delegates to Governor Howard and General Clarke at St. Louis, who stated that "when the Great Spirit gave the land to the Red Men, their ancestors, he foresaw that the White Men would come into the country and that the game would be destroyed; therefore, out of his great goodness he put lead into the ground that they, their wives and children might continue to exist" (Bradbury, 1904, p. 253; Lanman, 1856, vol. 1, p. 25). This was doubtless the principal source of the galena so abundant in certain mounds of the Mound Builders. There is also evidence that the Indians obtained galena from the outcrops of the southeast Missouri mines (Thwaites, 1904–07, vol. 26, p. 95). Probably the Tri-State district was known to the Indians for they informed Lieutenant Wilkinson "that the country to the northwest of the Osage village abounds with valuable lead mines" (Coues, 1895, vol. 2, p. 561). John S. Newberry (1892, p. 191) states that near Lexington, Ky., the Indians sank a trench over 300 feet long and from 10 to 12 feet deep on a galena-barite vein. He adds that "trees growing in the trench show it to be at least 500 years old." The Blue Bell Mine, Kootenai County, Idaho, is said to have been discovered by the whites owing to the fact that the Indians made bullets by smelting its galena (Laut, 1918, p. 89). It is by no means certain, however, that this particular discovery by Indians is very ancient. F. M. Endlich near Cook's Peak, N. Mex., in sinking a shaft broke into an old stope in which were stone tools and none of metal (Chapin, 1892, p. 30).

Galena, found as pebbles on the sea beach of Coronation Gulf, is used by the local Eskimo to blacken skins (Stefánsson, 1913, p. 443).

HEMATITE

Hematite was used by Eskimo, the Indians of Canada, many tribes in the United States, the Aztecs, Mayas, the old Panamanians, and Peruvians, and the Indians of Cuba, Venezuela, Bolivia, Ecuador, Brazil, and Argentina. Hematite was mined at Marquette, Mich., and at Iron Mountain, St. Francois County, and at Leslie, Franklin County, Mo. At Leslie, white miners have opened up an open cut in iron ore, 150 feet long, 100 feet wide, and from 15 to 20 feet deep. Honeycombing the area mined and extending beneath its deepest part are tortuous winzes, the work of Indians. Most of the openings are narrow and sinuous but some permit of a man standing up in them. Over 1,000 rude stone implements, all grooved and from 1 to 5 pounds in weight, were found in and around the workings. The material sought was soft hematite used as paint although solid hematite for implements and some flint were byproducts (Holmes, 1904, pp. 723–726). The Hopi got hematite for ceremonial pigment from Cataract Canyon, 110 miles west of the Hopi Reservation (Hough, 1902). The Peaux de Lièvre Indians got hematite near Fort Good Hope on the Mackenzie River. From its appearance they called it *Sa-ts-anne* or "bear's excrement" (Chambers, 1914, p. 284).

AZURITE AND MALACHITE

Malachite was used by the Indians of Arizona and New Mexico, by the Aztecs, Mayas, and Peruvians and by the Indians of British Columbia, eastern Canada, Ecuador, Colombia, Bolivia, and Argentina. Azurite was used by the Pueblos, Eskimo, the Mayas, and the Indians of Bolivia, eastern Canada, San Domingo, and Chile. According to Apache belief, a small bead of malachite attached to one's gun makes it shoot straight (Bourke, 1892, pp. 588–591).

The Pueblo Indians had many sources of these minerals. Apparently a squad from Oñate's expedition (1598) inspected a shaft three "estados" (about 16 feet) deep from which these minerals were obtained (Bolton, 1916, p. 244) either in the Aquarius or Hualpai Ranges. In Father Geronimo Zarate Salmeron's account of the same expedition, the blue used as paint in the Zuñi province and the green from Xémez, where "whole cargoes could be gathered," were also presumably oxidized copper ores (*in* Bolton, 1916, p. 269). The Hopi got malachite for ceremonial pigment from Cataract Canyon, 110 miles west of the Hopi Reservation (Hough, 1902). Copper carbonates were obtained for pigments by the Pueblos in the elevated region west of Luna, New Mex. (Hough, 1907, p. 59).

SMITHSONITE

Smithsonite was ornamentally used by the ancient Peruvians and probably by the Pueblos.

ATACAMITE AND BROCHANTITE

Beads of both these copper minerals have been found in Chile, near their source, the Chuquicamata mine. Atacamite was also used by the Peruvian and Argentina Indians, and brochantite by the Bolivian Indians.

CHRYSOCOLLA

Chrysocolla was popular among the old inhabitants of South America, artifacts of it being found in Peru, Bolivia, Chile, and Argentina. It was obtained at Quebrade de Cobres, northwestern Argentina, by the Diagüites. Among their workings is a 45° inclined shaft 30 meters deep   (Beuchat, 1912, p. 715). It was also used by Indians of California and Arizona.

PYRITE

Pyrite was rather widely used in both North and South America. It is not unusual for Maya pyrite mirrors and the firestones in Eskimo graves and those of the Maine Red Paint People to be altered to limonite, a possible scale of the rapidity of pyrite alteration.

In Labrador, northern Canada, and Alaska, the Eskimo used pyrite to strike fire as did the Indians of northwest and northern Canada and Newfoundland. It may be mentioned that the Aleutians to obtain fire strike together two flints rubbed in sulfur, the spark falling on lint powdered with sulfur which is obtained from the nearby volcanoes (Dall, 1870, p. 370). At Point Barrow, the Eskimo miners believe the pyrite, which occurs massive and as spherical concretions, to have fallen from the sky and hence it is called "firestone" (Ray, 1885, p. 46; Murdoch, 1892, p. 60; Hough, 1890, p. 574). Among the Cumberland Bay Eskimo snapping a whip with a piece of pyrite at the tip drives away evil spritis (Boas, 1907, p. 138). These people believe that some seals break a breathing hole through the ice with a stone held under the flipper. A hunter, if he kills such a seal, should, without looking at the stone, throw it over his shoulder, which changes it into pyrite and thereafter insures good luck in sealing (Boas, 1907, p. 152). The Iglulik Eskimo protect themselves from thunderstorms by laying out an amulet consisting of pyrite, a piece of white skin, and a small kamik sole (Rasmussen, *in* Weyer, 1932, p. 182, footnote). The Haneragmiut Eskimo procure pyrite for fire-making to the northwest of Coronation Bay and trade this to the Copper Eskimo (Stefánsson, 1919, pp. 74, p. 113).

It is common in the Mound Builder mounds of Ohio, that of the Muskingum Valley presumably coming from the adjoining hills where pyrite abounds. In 1826 an English traveler, Ash, found in a mound what from its luster he believed a large lump of gold. His laborers carefully covered up their work and secretly in a private room gave it the fire test. Their "gold" turned "black, filled the place with a sulfurous odor and then burst into 10,000 fragments" (Mitchener, 1876, pp. 24–25).

Pyrite was used extensively by Aztecs as inlays in their mosaics, as eyes for their statues, and, well polished, as mirrors. Zelia Nuttall (1901 a, p. 83) believes that pyrite mirrors were used in the sun-cult, to concentrate the rays of the sun and so light the sacred fire at noon on the days of the vernal equinox and summer solstice. Crushed marcasite was used as a face powder by certain Aztec priests (Bancroft, 1886, vol. 3, p. 340).

Don George Juan and Don de Ulloa (traveled, 1734) refer to ancient aboriginal mines of pyrite in Peru (Pinkerton, 1808–14, vol. 14, pp. 545–546). Pulverized pyrite was among the votive offerings to the gods.

The natives of Tierra del Fuego who, like the Eskimo, used iron pyrite for fire-making, obtained it from at least two mines known to us; one on the northern part of Tierra del Fuego Island and another near Mercury Sound, Clarence Island (Cooper, 1917, pp. 191–192). As its use to produce fire is noted as early as A. D. 1580, it was doubtless a pre-European custom (Lothrop, 1928, p. 64). The Fuegians prize it highly. Their neighbors, the Patagonians, not only used it for making fire but weighted the globular hide bags at the end of their bolas with it. Presumably they obtained it from the Fuegian country (Fitzroy, 1839, vol. 1, p. 62).

CANNEL COAL, JET, AND LIGNITE

Cannel coal, lignite, and jet were rather widely used by the North American Indians including the Aztecs and Mayas as well as by the Indians of Montserrat Island and Ecuador. These were used mainly as ornaments, although certain of the Pueblo people, the Assiniboin and other Northern Plains Indians, used it to a minor degree as fuel.

The Indians of Blennerhasset Island, W. Va., used to a considerable extent for pendants cannel coal which they procured locally (Hodge, 1922, p. 151).

Cannel coal adorned the northland in the Navaho Creation Legend. Jet occurs in Colorado and lignite widely in the Southwest, and their outcrops furnished abundant material to the Pueblo peoples. They mined lignite for fuel purposes 15 miles north of Holbrook, Ariz., and at Kokopuyama, northeastern Arizona (Hough, 1903, pp. 334, 335).

MICA

Mica was used by the Eskimo and the Indians of British Columbia, eastern Canada, the United States east of the Mississippi and California, by the Pueblos and Aztecs, and the Indians of Panama (paragonite), San Salvador, Peru, Ecuador, Venezuela, and Argentina.

Mica (largely muscovite) was mined by pits at many points in the Appalachian uplift from Alabama north-northeast to the St. Lawrence. It was an article of trade west as far as the Mississippi and south as far as Florida. Rock crystal and the more unusual pegmatitic minerals, often vividly colored, must have been occasional but prized byproducts. The Indians of North Carolina carried down some of their mica pits as far as surface weathering extended, that, is to ground-water level. Some of the pits were from 40 to 50 feet wide and from 75 to 100 feet long and, though in part filled up, are still from 15 to 20 feet deep. One North Carolina mica pit is 320 feet long and in places 30 feet deep (Smith, 1877, pp. 441–443). Old tunnels connecting pits from 50 to over 100 feet long are mentioned. They are from 3 to 3½ feet in height and much less in width. Large trees have grown within the pits. When the North Carolina mica boom was on in 1868–69, such pits were of value in relocating mica mines and the Indians proved to have been good prospectors (Kerr, vol. 1, 1875, pp. 300–301; 1880, p. 457). Incidentally the mountaineers were obsessed with the idea that the ancient pits were silver mines worked by De Soto's men (Foster, 1873, p. 370). D. B. Sterrett (1907, p. 401) found stone tools around some of the North Carolina mica mines and Wm. B. Phillips (1888, p. 662) states that the Indians used fire in breaking the rock. The latter reports that large pine trees (18 inches in diameter) have grown on the debris of the Alabama mica mines (Phillips, 1907, p. 671). Mica mines occur in 7 different counties in Alabama and at some 17 localities in North Carolina (Holmes, 1919, pp. 244–245). The old traveler Laudonniére (A. D. 1564) was shown by Indians large mica plates found in the Appalachians with "christal" and "slate stone." Ralph Lane (A. D. 1585–86) heard also of a "marvelous and strange mineral" occurring in large plates, which was mined to the west of Roanoke (Packard, 1893, pp. 162–163). In Amelia County, Va., there are aboriginal mica pits 12 feet deep (Fontaine, 1883, pp. 330–339). Fire was used to break the rock.

Mica was commonly used by the Mound Builders, and J. Priest (1838, p. 179) mentions one piece, 3 feet long, 1½ feet wide, and 1½ inches thick, a fair sized "book." From a single mound, as many as 250 mica objects and as much as 20 bushels of mica have been reclaimed. It doubtless came from the Appalachian mines. "Synthetic" pearls were made by the Mound Builders by wrapping a coat of mica around wooden beads (Davis, 1931, p. 136). Again, beads of clay were covered with crushed mica (Lilly, 1937, p. 210).

Among the Delaware Indians, mica laminae are placed in medicine bags and are powerful "rain medicine" as they are believed to be scales of the great mythological Horned Serpent. Merely place a few "scales" on a rock beside some stream and the black thunder clouds gather and refresh the thirsty cornfields with rain (Harrington, 1913, p. 226).

Fuchsite beads are reported from an ancient Guatemalan grave (Bauer, 1900, p. 291).

LABRADORITE

Labradorite, which was introduced to the scientific world by Moravian missionaries in 1770, was presumably procured by them from the Eskimo, who still bring fine specimens from the interior of Labrador (Packard, 1891, pp. 272–283). They know it as "the fire-rock" (Browne, 1909, p. 155). Captain Cartwright, who was in Labrador from 1770–86 (Cartwright, 1911, p. 347), mentions it being picked up by the Eskimo. Eskimo chiefs used it ornamentally over 50 years ago (Tuttle, 1885, p. 65).

Anorthite was used in Panama.

SUNSTONE

Dr. H. P. Wightman states that the Apache Indians collected sunstone (variety *andesine*) from their reservation not far from Globe, Ariz. (Sterrett, 1916, p. 322).

MOONSTONE

Moonstone was among the gems excavated at Ticoman, Mexico, in a grave of Toltec or pre-Toltec culture (Vaillant, 1930, pp. 610–616).

AMAZONSTONE

Amazonstone was used by the Aztecs, Mayas, and the Indians of Wisconsin, California, Trinidad, Venezuela, and Brazil.

SLATE

The Haidahs, skillful carvers in slate, obtained the raw product from a quarry on Queen Charlotte's Island.

VARISCITE

Mr. Don Maguire, according to Dr. G. F. Kunz, reports that in the vicinity of the variscite locality near old Camp Floyd in Cedar Valley, Utah, artifacts and rock inscriptions are common. No old workings have as yet been found, however, but variscite was used by the ancient Pueblo people of the region.

CALAMINE

Calamine of predominant blue color, but in part gray and green, is possessed by the Yaquis of Chihuahua, Mexico. They use it as votive offerings, believing it has magical qualities (Sterrett, 1909, p. 812).

FLUORSPAR

Fluorspar was used as an ornamental stone by the Indians of Missouri, Illinois, Tennessee, Kentucky, Indiana, and California, the Pueblos, Aztecs, and Peruvians, and the Bolivian Indians.

The fluorite used by the Mound Builders was probably picked up from the outcrop as E. C. Clark[6] knows of no aboriginal pits in the Illinois-Kentucky field. He states that most of the fluorite employed by the Mound Builders apparently came from the Illinois part of the field.

AMBER

Amber was commonly used by the Eskimo and the Indians of Alaska and British Columbia, the Aztecs and Mayas, the Peruvians and the Indians of Santo Domingo and the Lesser Antilles and Colombia. A fossil gum was also used by the Brazilian Indians and a fossil resin closely resembling amber has been found in Mound Builder mounds in Ross County, Ohio. Reported occurrences of amber artifacts in Virginia and Tennessee may or may not be authentic.

The Eskimo of Point Barrow find from time to time amber on the beach and use it rough as amulets and rarely cut it into beads. It is called aúma, i. e., "a live coal," a descriptive figure of speech (Murdoch, 1892, p. 61). Ernest de K. Leffingwell (1919, p. 179) saw the Point Barrow natives "pick up a few pieces [of amber] a quarter of an inch in diameter from the protected beaches between Harrison and Smith Bays." The Eskimo also got amber for beads from the alluvium of the Yukon delta and from the Tertiary formations of the Fox Islands (Holmes, 1907, p. 48). The Koniagas of Kodiak Island prize labrets, ear ornaments, and pendants of amber which at times, particularly after earthquakes, is said to be thrown up upon the south side of the island. Broken beads and pieces of amber are placed on the graves of the wealthy. It is also an important, though rare and costly, article of commerce among them. (Cox, 1787, p. 212; Bancroft, 1886, vol. 1, pp. 72–73; Dall, 1870, p. 403; Petroff, 1884, p. 138.) That amber was widely traded in among the Eskimo long ago is shown by the presence of beads and uncut lumps in Thule culture ruins (about A. D. 600–1600) at Naujan on the shore of Fox Channel. Amber beads said to be of Asian origin were found in an Aleutian grave on Unalaska Island (Weyer, Jr., 1929, p. 234).

---

[6] Personal communication.

Gerard Fowke (1894, p. 16) found in a grave in Rockbridge County, Va., a bead "resembling amber," which, he suggests, possibly may be of European origin. The occasional finding of amber on the adjacent Virginian coast appears to render such a conclusion unnecessary. W. M. Clark (1878, p. 275) reported amber beads in a Tennessee mound but as these were stolen before being placed in a museum, the observation must be accepted with reservations.

Amber and labrets of amber were among the tribute to be paid Montezuma by certain of the districts of Mexico, particularly the cities on the Atlantic coast and of Chiapas, a present-day locality. In Aztec times it was an important article of commerce. Sahagun (1880, p. 771) reports that amber was obtained by the Aztecs from "mines in the mountainous country." Clavigero reports that it was used as an ornament mounted in gold.

The amber of the north coast of Santo Domingo was gathered in pre-Columbian days. It was the first gem material recognized in the New World by the whites, as Christopher Columbus, in his account of his second voyage, says that the island contains "mines of copper, azure, and amber" (Kerr, 1811, vol. 3, p. 131).

Soapstone (Steatite)

Soapstone was used by practically all the North and Central American peoples, by the Indians of Puerto Rico, Venezuela, Colombia, Bolivia, Brazil, Argentina, and Chile, and by the Peruvians. The Cherokees of the Great Smokies, N. C., still make soapstone pipes, largely, however, for tourist consumption (Morley, 1913, p. 238) and Peter Kalm states that in 1748 soapstone pots were still used among the Delawares.

Steatite was quarried in a large number of places in the Appalachian uplift from Newfoundland to Alabama, some 33 being known to the writer. Other localities were Wyoming and Lac de la Pluie, southwest of Lake of Woods, Ontario (Mackenzie, 1902, vol. 1, p. xcii). Soapstone was also obtained in the Jacumba region, San Diego County, Calif. (Gifford, 1931, p. 29), and on Santa Catalina Island (Kroeber, 1925, p. 629). Stone hammers, mauls, and picks are common at such localities. A peculiar feature of the quarrying is the fact that bowls were in some instances largely shaped in place, then undercut, and only then broken off from the solid rock by gradual pressure of the chisel around the base of the bowl (Schumacher, 1879, vol. 7, pp. 117–121). At least the pots and other artifacts were usually roughly hewn at the quarry although they may have been finished at the home village. A. J. Pickett (1851, vol. 1, p. 177) states, from Indian testimony now over 100 years old, that in Alabama the Indians "cut out the pieces with flint rocks fixed in wooden handles. After working around as deep as they desired, the piece was pried out of the rock."

In Norway and Sweden in the Viking time, about 1,200 years ago, the Scandinavians cut pots from soapstone in place much as did the American Indians of that or later times (Grieg, 1930; 1932, pp. 88–106). The Laplanders of northwestern Sweden [7] also cut their pots directly from the rock outcrop.

At Johnston, R. I., the largest pits are 10 feet long, 6 feet wide, and are now 5 feet deep although originally doubtless 15 feet deep (Denison, *in* Chase, 1885, pp. 900–901). The Narragansett Indians were famous steatite artisans and traders, and their pipes made of local steatite were in demand not only among the Mohawk but also by "our English Tobacconists for their rarity, strength, handsomenesse and coolnesse" (Wood, 1634, p. 65).

The Cumberland Sound Eskimo when breaking steatite from a quarry, "deposit a trifling present at the place, because otherwise the stone would become hard" (Boas, 1907, p. 138). The Eskimo on the west coast of Hudson Bay and the Copper Eskimo believe that steatite should not be worked while the people are living on the ice (Boas, 1907, p. 149; Jenness, 1922, p. 184). The former sometimes use steatite as bullets when lead is scarce. It is mined by Eskimo at the mouth of Tree River, which flows into Coronation Gulf, 75 miles east of the Coppermine mouth. The Coppermine Eskimo are dealers in soapstone lamps and pots, and at many of the soapstone localities the main occupation of certain Eskimo is pot making. Families from as far away as Cape Bexley visit the Tree River for the stone, being en route 1, 2, or more years and such trips are the subject of local songs. It is also distributed by tribal barter. It is believed by Stefánsson that this and localities east of it once supplied soapstone cooking utensils as far west as Siberia (Stefánsson, 1919, pp. 68, 112–113; 1914, p. 25). Stefánsson (1919, p. 28) states that to make a pot takes all an Eskimo's spare time for a year and that certain of the more skillful members of the tribe specialize in making such utensils.

CATLINITE

Catlinite (*Eyanskah* in Sioux) (Neill, 1858, p. 515) occurs in the Coteau des Prairies, Minn., on the Red Cedar branch of the Chippewa River, Wis. (Schoolcraft, 1853, p. 206; Strong et al., 1882, vol. 4, pt. 5, p. 578; West, 1934, pp. 72–73, 330–331; De la Ronde, 1876, pp. 348–349; Barrett, 1926; West, 1910, pp. 31–34; Brown, 1914, vol. 13, pp. 75, 80–82), in Scioto County, Ohio (Shetrone, 1930, p. 178), and in Arizona. The latter is probably the oldest of the localities, as catlinite artifacts occur in the Pueblo II culture stage of Utah (A. D. 200–800). The next oldest of the localities certainly worked by the aborigines was the Ohio locality. Its trade area was

---

[7] Personal communication, Hans Lundberg.

much more restricted than that of the Minnesota catlinite, but it was extensively used in Ohio and Kentucky and in some instances reached even Iowa and Wisconsin. The Minnesota and Wisconsin localities, while probably the youngest of the catlinite mines worked by the aborigines, are well over 300 years old. Sioux myths connecting catlinite with the creation of man suggest a greater age, but myths of the long ago conceivably can grow in a day. Catlinite was also obtained from the glacial drifts in the upper Mississippi Valley (pl. 5).

To the Coteau des Prairies, the surrounding tribes from hundreds of miles around made yearly mining pilgrimages to obtain material for their pipes. The Great Spirit, after miraculously forming the pipestone, had dedicated the ground as neutral property where war was taboo, an admonition for a time respected. Even while en route to the quarries, the Indians' bitterest enemies would not attack them. (For a poetical rendering of Indian catlinite myths, see Longfellow's Hiawatha.) According to the Sioux, catlinite attained its color by being stained by the blood of buffalo slain by the Great Spirit, while to the Indians of the upper Missouri, it was the flesh of Indians drowned in a great flood (Armstrong, 1901, pp. 2–4, 11).

L. N. Nicollet, who visited the quarry in the Coteau des Prairies in 1838–39 (1843, pp. 15–17), adds that the Indians believe that when they visit the quarry they are always saluted by lightning and thunder and that its discovery was due to a deep path worn down into the catlinite bed by the buffalo, the path being still visible at the time of Nicollet's visit. Three days of purification preceded the Indians' visit to the quarry during which time he who was to do the quarrying must be continent. The Abbé Domenech (1860, vol. 2, p. 347) adds that during this period the miners fasted. Provided the pit, which the Indian miner sinks, does not encounter catlinite of good quality, he is considered to have "impudently boasted of his purity. He is compelled to retire: and another takes his place." A Sioux who visited the quarry about that time says that first there was a feast to the spirits of the place and then before quarrying a religious dance was held (Dodge, 1877, p. XLVII). The Indians (Domenech, 1860, vol. 2, p. 273) were loath to have white men visit the quarry as their presence was a profanation which would draw down the wrath of heaven on the Indians. Before mining began the medicine men invoked and propitiated the spirits of two glacial boulders nearby, symbolizing two squaws, the guardian spirits of the place (Schoolcraft, 1851–57: Catlin, 1913, vol. 1, pp. 25–26; vol. 2, pp. 186–195, 228–234).

A small creek at the foot of a quartzite ridge probably originally exposed the thin bed (18 inches) of reddish fine-grained somewhat metamorphosed clayey sediment (catlinite). At the base of the wall and parallel to it is a prairie one-half mile wide. On it for a distance of almost a mile are many pits from 20 to 40 feet wide and from 4 to

10 feet deep sunk through the soil to procure the pipestone (pl. 5). The tools were roughly shaped hammers and sledges from the nearby quartzite ledges, some of the hammers being grooved. Hieroglyphics are common on the faces of the rock ledge and tradition says that each Indian before venturing to quarry Catlinite inscribed his totem thereon (Mallery, 1893, p. 87). Remains of ancient camp sites abound nearby and in them are found partially worked fragments of pipestone, the material in part having been carved at the quarry, perhaps because with age this rock is said by some to become indurated. (Holmes, 1892, p. 277; 1919, pt. 1, pp. 109, 253–263; Nicollet, 1843, p. 15; Winchell, 1884, pp. 541–542; Hayden, 1867, pp. 19–22; West, 1934, p. 329; Lynd, 1865.)

Groseilliers and Radisson in A. D. 1658–60 were perhaps the first whites to see catlinite for they mention pipes of a red stone owned by the Nation of Beef living west of Lake Superior. Father Marquette (Repplier, 1929, p. 151) smoked a pipe of peace of catlinite in 1673 and Le Seur in 1699–1700 mentions the Hinhoneton's "village of the red stone quarry" (Shea, 1861, p. 111). Father Charlevoix (1763) mentions its source as among the "Ajouez (Iowa) beyond the Mississippi." Initial quarrying by Indians must have preceded the seventeenth century. Le Page du Pratz (1758, vol. 1, p. 326) states that when M. de Bourgmont, in 1724, was en route to visit the Padoucas, he saw a cliff on the banks of the Missouri, consisting of a "mass of red stone with white spots like porphyry" but soft, easily worked into pipes and fire-resistant; this lay between two valueless stones. "The Indians of the country have contrived to strike off pieces thereof with their arrows and after they fall in the water plunge in for them." This strangely perverted account of the method of obtaining catlinite is similar to that by which Pliny states the Persians obtained turquoise.

Jonathan Carver (1778, p. 99), who traveled in the upper Mississippi region in 1766–68, shows on his map Couteau des Prairies as "Country of Peace." Lewis and Clarke (1804–06) also speak of the Indian tribes meeting there in "friendship" to collect stones for pipes (Coues, 1893, vol. 1, p. 80) but in 1837 when George Catlin visited it (1848, vol. 2, p. 166) the Sioux were in possession and did not permit their enemies to procure pipestone. Indeed an old chief of the Sauk complained to Catlin that, while as a young man, he visited the place to dig catlinite "now their pipes were old and few." "The Dakotans have spilled the blood of the red men on that place and the Great Spirit is offended" (Catlin, 1848, vol. 2, p. 171). Thus, prior to about 1810 the quarry was neutral ground, but after that date was in the possession of the Sioux. About 60 years ago, Professor Crane reported that 300 Yankton Sioux took part in the annual pilgrimage to the quarry, and an Indian chief claimed 100 years ago he had seen 6,000 Indians encamped at the quarry for 2 months (Barber, 1883, pp.

745–765). The Yankton Sioux by article 8 of the treaty with the United States, dated April 19, 1858, have the right in perpetuity to mine catlinite within about a square mile surrounding the quarry (S. Dak. Hist. Coll., 1902, vol. 1, app., p. 449). While these people and their friends, the Flandreau Sioux, still visit [8] the locality, many of the pipes and trinkets sold since at least 1866 have been of white manufacture. Indeed, Indian traders had glass beads, imitating catlinite, by the end of the eighteenth century.

The trade in catlinite was Nation-wide, extending from Canada to the Gulf of Mexico and from the Atlantic to well within the Rocky Mountains Iowa Indians, according to Father Louis André (Thwaites, 1896–1901, vol. 60, p. 203), in 1676, had Minnesota catlinite and the Iroquois and Algonquin peoples on the Atlantic coast by intertribal barter late in the seventeenth century. Peter Kalm, Professor of Economy, University of Abo, Finland, who traveled in North America from 1748 to 1751 (Pinkerton, 1808–14, vol. 13, p. 516; Kalm, 1772, vol. 2, p. 43) says the chiefs of the Indians of Pennsylvania had pipes ingeniously made of "very fine red pot-stone or a kind of serpentine marble of the kind which Father Charlevoix says comes from beyond the Mississippi." They were very scarce and were valued "as much as a piece of silver of the same size and sometimes they make it still dearer." Loskiel (1789, p. 66) says that the Delawares and Iroquois got their pipes direct from Indians "who live near the Marble River, on the western side of the Mississippi, where they extract it from a mountain." If so, these Indian merchants carried their wares 1,000 miles from their homes. Among the Wisconsin Menomini, the journey to the quarry in Minnesota being long, small blocks of catlinite were locally valued at $100. At Fort Union in 1852, Kurz purchased a "charming" catlinite pipe from an Absaroka for $7, although among the Crows it would have been worth a packhorse (Kurz, 1937, p. 257). He adds that, though fashioned by the Sioux, they are articles of barter among all the other tribes.

The red pipestone found in the Ohio Mound Builders mounds was doubtless largely obtained from the Ohio pits (Shetrone, 1930, p. 178, and personal communication) but that from the mounds in Lyon County, Iowa, presumably, as A. R. Fulton holds (1882, p. 89), came from the Minnesota quarry. The Ohio pipestone is of lighter color than that of the other localities, being pinkish or grayish.

There are a number of catlinite quarries in Barron (at least five quarries) and Chippewa Counties, Wis. The largest quarry, situated on a hill, is about 25 feet square and not over 3 feet deep. In places, the indurated shale was stoped back beneath the overlying quartzite. The Indians worked here at least as late as 1914. The catlinite is reddish brown in color and was used extensively by the Indians. De la

---

[8] Personal communication, G. L. Chesley, postmaster, Pipestone, Minn.

Ronde visited the locality over 100 years ago and states that the Indians travel many miles to obtain the catlinite and that while there peace reigns. He tells a dramatic story of a Sioux and a Chippewa who by chance met at the mine and peaceably procured their pipe material but as soon as they were a respectable distance from the sacred spot fought a duel to the death.

The Pueblos imitated catlinite in pottery, showing the esteem in which they held catlinite, some of which occurs in the ruins of Arizona. It doubtless came from the Jerome Junction, Arizona locality (Schrader, Stone, and Sanford, 1917, p. 18).

OBSIDIAN

Obsidian was used throughout North and Central America, except in eastern Canada, New England, the North Atlantic (a single obsidian artifact has been found in Pennsylvania), and the southeastern States (artifacts have been reported from Georgia and Alabama, however) and it was used in South America by the Indians of Colombia, Ecuador, Bolivia, British Guiana, Peru, Argentina, Chile, and Brazil, and also by those of the Lesser Antilles. By trade it traveled vast distances.

Obsidian was quarried at Obsidian Cliff, Yellowstone National Park, and elsewhere in the Yellowstone and Snake River Valleys as well as in Utah, New Mexico, Arizona, and Nevada (Holmes, 1919, pp. 214–227). Apparently the Obsidian Cliff, Yellowstone National Park, was neutral ground to Indians seeking arrowhead material (Alter, 1925, p. 381). From shallow shafts, drifts were driven (Brower, 1897, pp. 20–24). Shell heaps around San Francisco Bay, in the opinion of A. L. Kroeber (1925, pp. 927–930), show that obsidian was used from 3,000 to 3,500 years ago. Unworked blocks are buried therein as if of great value. It is reported in Minnesota in deposits antedating the last glacial state (Hagie, 1936).

Obsidian was obtained by the aborigines in some 10 California localities either in place or as pebbles. The Pit River Indians made long trips to Sugar Hill in the summer to procure obsidian pebbles (Kniffen, 1928). The hill was sacred and the Indians feared to offend its spirit. The Klamath Indians of southeastern Oregon believed that arrows made from obsidian obtained on a mountain west of Klamath Lake were poisonous (Spier, 1930, p. 32; Rust, vol. 7, 1905, pp. 688–695). Herbert J. Spinden (1908, vol. 2) states the Nez Percé's name for the John Day River meant "obsidian river." The Mandan Indians remelted glass and cast characteristic beads: if, as tradition holds, they learned the art from the Snake Indians, their original raw material may have been obsidian from the Yellowstone National Park, rather than clay (Matthews, 1877, pp. 22–23).

Obsidian was abundantly used by the Mound Builders and the Hopewell (Ohio) people had ceremonial knives 18 inches long and over 6 inches wide. Evidently blocks of obsidian were imported and the material worked up in the Mound Builders' village. Shetrone (1930, p. 65) thinks that, as obsidian and grizzly bear teeth are found together in the Hopewell culture mounds and not in quantity, at least, to the west, the Hopewell men sent special expeditions to the Yellowstone Park to obtain obsidian.

The pueblos of New Mexico and Arizona had obsidian at hand at many places. They not only quarried it but also apparently depended for their supply on the shattered outcrops. The Tewa Indians of the upper Rio Grande Valley, N. Mex., believe that the flaking of the outcrop is due to lightning striking it (Harrington, 1916, p. 59).

Obsidian was doubtless used a very long time ago by the peoples of the Southwest, as Earl H. Morris (1919, p. 202) reports that the obsidian implements from ancient ruins in southwestern Colorado are so old that they have acquired a "dull gloss," or patina. That it was long in use among certain of the Indians is also indicated by the Athabascan folk tale in which it was one of the four substances existing before the world was created (Goddard, 1827, p. 180).

The Aztecs and their predecessors used obsidian extensively, fashioning from it spear and arrow points, knives, razors, and swords, mirrors, beautiful masks, and dainty ear ornaments. The Aztecs called it *iztli* and because of its many uses it was surnamed *teotetl* (divine stone) (Bancroft, 1883, vol. 3, p. 238). So abundantly was it used by the Aztecs that some of the refuse heaps around Mexico City are black with its fragments (Holmes, 1900, p. 406). Obsidian must have been long known to the Aztecs as one of their gods was Itzpopalotl (Obsidian Butterfly) (Verrill, 1929, p. 185). Further at Cuichuilco near Mexico City, Dean Byron Cummings in 1916 found a structure which, in its relation to certain lava flows, he believed long antedated the Aztecs. Nearby, Mrs. Zelia Nuttal found fragments of obsidian flakes "with a dull surface and a patina which unquestionably indicate great antiquity" in an ancient river bed, 17 feet beneath the lava bed (Mason, 1931, p. 30).

Alexander de Humboldt rediscovered the old Aztec obsidian pits of Sierra de las Navajas some 10 miles east of Pachuca (1811, vol. 3, p. 122; 1815, vol. 1, p. 337). The aborigines, over hundreds of acres, according to W. H. Holmes (1900, vol. 2, pp. 405–416) in an area 1½ miles long and in places one-half wide, sunk shallow pits and gophered out horizontally from them where the obsidian was of good quality. The spall heaps nearby run into the hundreds of tons. W. H. Holmes, who says that the deepest pits are now 20 feet deep although once deeper, estimates that the workings are about as ex-

KEY TO ABBREVIATIONS USED IN TABLE 1

Central America:
    Costa Rica_____ C.
    Nicaragua_____ N.
    Panama_____ P.
    San Salvador_____ S.

West Indies:
    Santo Domingo (Hispaniola)_ S.
    Montserrat_____ M.
    Jamaica_____ J.
    Puerto Rico_____ P.
    Cuba_____ C.
    Guadeloupe_____ G.
    Trinidad_____ T.
    Lesser Antilles_____ L. A.

Northern South America:
    British Guiana_____ Br.
    Venezuela_____ V.
    Colombia_____ C.
    Ecuador_____ E.
    Bolivia (in part Incan)_____ B.
    Dutch Guiana_____ D. G.

Southern South America:
    Argentina_____ A.
    Chile_____ C.
    Paraguay_____ P.
    Brazil_____ B.

tensive as the great flint quarries at Hot Springs, Ark., and Flint Ridge, Ohio. Stone hammers, discoidal or globular in form, were found near the pits. In part, at least, the material was worked locally but it was also transported in the rough to distant markets. Löwenstern (1843, pp. 244–245) describes these mines as trenches from 1 to 2 meters wide and of varying depth. From the extent of the pits it is believed that mining began some centuries before the arrival of the Spaniards (Tylor, 1861). It was also quarried at Zacaultipan and 15 miles south of Tulancingo, Hidalgo; at Teuchitlán, Ixtlan de los Buenos Aires, and Etzatlan, Jalisco; Pica de Orizabo, Veracruz; Zinapécuaro, Michoacan; and elsewhere in Mexico. At Teuchitlán some of the flakes are so very old that they are covered with a thick white crust (Breton, 1905, pp. 265–268).

The Mayas got their obsidian doubtless in part, at least, from the extensive ancient quarries at La Joya, 18 miles east of Guatemala City. It also occurs at Fiscal (on the railroad from Guatemala to Zacapa) and near Antiqua (Holmes, 1919). At Flores, Guatemala, the local "small change" consists of oblong pieces of obsidian, the value depending on the size and shape of the piece (Boddam-Whetham, 1877, p. 296).

Don George Juan and Don de Ulloa (who travelled in Peru in 1734) refer to ancient aboriginal mines of obsidian in that country (Pinkerton, vol. 14, pp. 545–546). The Ecuador Indians had an obsidian quarry at Guamani and a thriving commerce in it was carried on with the coastal Indians. In their Quichua language it was known as *aya-collqui*, "silver of the dead," certainly an apt poetic name.

OTHER MINERALS

Diopside was used by the Indians of British Columbia and the Pueblos, as was actinolite by the California Indians and the Pueblos. Scapolite was utilized by the Pueblos and the Peruvian Indians. Sillimanite was used by the Pueblos and the Brazilian Indians. Saussurite was employed by the Indians of Ecuador. Chlorite was commonly used in North America, only rarely in South America. Agalmatolite was made into artifacts by the Mayas and the Indians of Ecuador, Colombia, and Chile. Apatite was known by the Pueblos. Barite was used by the Indians of Kentucky and by those of Georgia. Artifacts are common in the vicinity of Cartersville in the latter State, a district in which barite residuals are abundant. Celestite was made use of by the Indians of New York. Gypsum was utilized widely by the Eskimo, the Indians of Eastern Canada and the northeastern United States, those of the upper Mississippi Valley, northern plains, the Rocky Mountain States and California; also by the Pueblos, the northern Mexicans, the Mayas, and the Indians of Costa Rica, Cuba, Bahamas, Chile, and Argentina.

LIST OF MINES OPERATED BY THE INDIANS (pls. 4, 5)

Aga,   Agalmatolite:
      Paccha, Ecuador.

A,   Agate:
      Southampton Island, Canada; AgateBay, Two Harbor, Minn.; Spedes Valley, Wash. (also opal, etc.); Millers Island, Wash. (also opal, etc.); cataracts, Demarara R., British Guiana; Parima Mountains, Venezuela.

Ag,   Agatized wood:
      Petrified Forest, Adamana, Ariz. (also amethyst, smoky quartz, etc.); Petrified Forest, 25 miles south of Chambers, Ariz. (also chalcedony, etc.).

Al,   Alabaster:
      Wyandotte Cave, Crawford County, Ind.; Tecali, Puebla, Mexico.

Am,   Amber:
      Aliaska Peninsula; Aliaka Isle; Unalaska Island; Umnak Island, Yukon Delta, Alaska; Ookamak Island and south side Kodiak Island, Alaska; Point Barrow, Alaska; north coast, Santo Domingo; Chiapas, Mexico.

At,   Atacamite:
      Chuquicamata, Chile.

Ba,   Barite:
      Bartow County, Ga.

B,   Brochantite:
      Chuquicamata, Chile; Corocoro, Bolivia.

Ca,   Cannel coal:
      Blennerhasset Island, W. Va.

Ct.,   Catlinite:
      Pipestone, Minn.; Chippewa County, Wis.; Barron County, Wis., five quarries; Scioto County, Ohio.

C,   Chalcedony:
      Warsaw, Coshocton County, Ohio (numerous quarries); Redondo Beach, Calif.; Ballast Point, 5 miles below Tampa, Fla.; Wagon Wheel Gap, Colo. (also jasper); Saugus Center, Mass.; 100 miles northwest of Pinto Basin, Calif.

Ch,   Chrysocolla:
      Quebrade de Cobres, northwestern Argentina.

Cy,   Chrysoprase:
      Tulare County, Calif. (possible).

E,   Emerald:
      Coscuez, Chivor-Somondoco, and Muzo, Colombia.

F,   Fluorite:
      Southern Illinois.

G,   Galena:
      Golovin Bay, Alaska; Coronation Bay, Canada; Anse-a-la Mine, Quebec; Wisconsin-Iowa-Illinois lead district; southeastern Missouri; Cook's Peak, N. Mex.; near Lexington, Ky.

Ga,   Garnet:
      Picuries, Taos County, N. Mex.; Navaho Reservation, N. Mex.; northeast Arizona; north central Arizona; Jaco Lake, Chihuahua, Mexico; near Quito, Ecuador.

H,   Hematite and red ochre:

> Mackenzie River at Fort Good Hope, Canada; Tulameen River, British Columbia; Tanana River, Alaska; Leslie, Franklin County, Mo.; Iron Mountain, Mo.; Marquette, Mich.; 44° N., 111° W., Ross Co., Ohio; Cataract Canyon, Ariz.; Red Canyon, Green River, Utah; Wellington, Calif.; Manzano Mountains, N. Mex.; 4 miles southwest of Zuñi, N. Mex.; Katahdin, Maine; Nodules, Kanawha Valley, W. Va.; Kaimak, Argentina.

I,   Iron pyrite:

> Point Barrow, Alaska; Rowsell Harbour, Labrador; northwest of Coronation Bay, northern Canada; 92°30′ W., 69°30′ N.; Victoria Island; creek east of Coppermine River; near Wager Inlet and Repulse Bay; Bad Creek, 70° N., 117° W.; Mackenzie 10 miles below Fort Good Hope, alluvial; Whitemud River, Saskatchewan; north part Tierra del Fuego Island; Mercury Sound, Clarence Island.

J,   Jasper:

> Saugus Center, Mass.; Chester Bucks, Lancaster, Lehigh, and Berks Counties, Pa.; Flint Ridge, Licking, and Muskingum Counties, Ohio (also chalcedony); Normanskill on Hudson River, N. Y.; Converse County, Wyo. (also chalcedony); Delaware River, Mercer County, N. J.; St. Tammany Parish, La.; 40 miles south of Twenty-nine Palms, Calif.; 40 miles north of Pinto Basin, Calif.; Calpulalpam, Mexico; Mount Roraima, British Guiana.

La,   Labradorite:

> Labrador, near Paul's Island.

L,   Lignite:

> Fifteen miles north of Holbrook, Kokopuyama, and Tusayan, Ariz.

Li,   Limestone:

> For pipes, Ottawa River at Portage du Grand Calumet; Lake Winnipeg River to west of Pike River; Falls of Montmorenci near Quebec; Sitka, Alaska; Lynn Canal, Alaska; Nipigon Island, Lake Superior; Flint River, Ga.; St. Joseph's Island, Ontario.

Mg,   Magnesite:

> White Buttes, near Cache Creek, Calif.; Sulphur Bank, Clear Lake, Calif.; Kaolin, Nev.

Ma,   Malachite:

> Cataract Canyon, Ariz.; Azurite Mountains, N. Mex.

Ma, Az,   Malachite and azurite:

> Highland, west of Luna, New Mex.; Aquarius and Hualpai Ranges, Ariz.

Mo,   Moss agate:

> Between Fort Ellis and Yellowstone River, Mont.; valley north of Uinta Mountains and San Rafael Valley, Utah.

M,   Mica:

> Spruce Pine (two pits), Bandana (two pits, Mitchell County), Bakersville, (Mitchell County), Yancey County, and Macon County (12 pits), N. C.; Amelia County, Va.; Chilton County, Jefferson County, Coosa County, Clay County, Randolph County, and Cleburne County, Ala.

N,   Nephrite:

> Jade Mountain, north side of Kobuk River, Alaska; Fraser River and Thompson River, British Columbia; Puget Sound; divide headwaters of Orinoco and Rio Branco Railroad, British Guiana.

●,   Obsidian:

    Rocky Mountains, about 62°30′ N.; Mount Anahim, 100 miles north-
east Prince Rupert, British Columbia; Mountains of Thompson
River, British Columbia; Klamath Lake, Oreg.; John Day River,
Oreg.; Glass Butte, Lake County, Oreg.; Obsidian Cliff, Yellow-
stone National Park; Promontory Point, Great Salt Lake, near
Willard; Shingle Spring (Sierra Nevadas), Clear Lake, lower Clear
Lake, Head of Napa Valley, Upper Sonomo Valley, Cole Creek,
Shasta County, Glass Mountain, Sugar Hill, Wheatland, Cortina,
Mount Kilili (near Tuolumne), Mount Kanaktai (Sonoma County),
and near Fillmore (Ventura County), Calif; near Pecos Church,
headwaters Gila River, and Jemez Plateau, N. Mex.; east of Silver
Peak, Nev.; Robinson Crater and Mount Sitgreaves (latter detrital)
in San Francisco Mountains, Ariz.; Sierra de las Navajas, Zacaultipan,
and south of Tulancingo (Hidalgo), Cerro Tepayo, Teuchitlan and
Ixtlan de los Buenos Aires and Etzatlan (Jalisco); Pica de Orizabo
(Veracruz), and Zinapecuaro (Michoacan), Mexico; La Joya,
Antiqua, and Fiscal, Guatemala; Patagonia (detrital 48°10′ S.,
72° W.); Guamani, Ecuador.

O   Olivine:

    Navaho Reservation, N. Mex.

R,   Rock crystal:

    Mackenzie River mouth, Canada; James River, 12 miles above
Richmond, Va.; Hot Springs, Ark.; Little Falls, Morrison County,
Minn.; west end Wichita Mountains, Okla. (also jasper); Yakima
Valley, Wash.; Armonk, Westchester County, N. Y.; east end,
Long Island, N. Y.; Tiquie River, Colombia; near Guayaquil,
Ecuador; Pikin Mountain, Brazil; Manhattan Island, N. Y.;
Compounce, Conn.; northern Oaxaca, Mexico.

S,   Selenite:

    Southern Nebraska; near Zuñi, N. Mex.; southeast New Mexico.; near
Santa Fe, N. Mex.; 44°10′ N., 104°18′ W., S. Dak.; Gypsum Cave,
near Las Vegas, Nev.; Kaolin, near St. Thomas, Nev.; Mammoth
Cave, Ky.

Se,   Serpentine:

    Two to two and one-half miles north of Phillipsburg, N. J.; Red Rock,
Grant County, N. Mex.; streams near Jade Mountain, Alaska;
Pipestone Lake, 54°30′ N., 93°30′ W.; Anderson Lake, British
Columbia.

Sl,   Slate:

    Skidegateon, Queen Charlotte Island, British Columbia; 5 miles
up Mattawa River; Elk River, Canada.

▲,   Soapstone:

    Ukasiksalik and Nachvok Fiord, (several localities), Labrador; Buck
River, 95° W., 67° N.; Simikameen R., British Columbia; Pipe-
stone Lake, Manitoba; Lac de la Pluie, Canada; Reindeer Lake,
west shore 103° W., 57°30′ N.; just west of Great Slave Lake;
Utkusikaluk about 111° W., 67°40′ N.; 68° 10′ N., 114° W., on
Coronation Bay; 112°30′ W., 67°45′ N.; Reindeer Island in Great
Slave Lake; Lake of Woods, Akkoolee, near Repulse Bay; 96° W.,
66°30′ N.; 90° W., south of Pelly Bay; Arctic Ocean, 60 miles east
of Mackenzie River; Cumberland Sound; mouth, Tree River;
Fleur de Lis, Newfoundland; Johnson, Vt.; Westfield, North
Wilbraham, and Millbury, Mass.; Johnston and Providence, R. I.;

Bristol, Nepaug, Portland, and Harwinton, Conn.; Christiana and Bald Hill, Lancaster County, Pa.; four localities Patuxent Valley (Montgomery and Howard Counties), Olney, and Clifton, Md.; Washington, D. C.; below Little Falls, near Washington, D. C.; Culpeper, Wayland Mill (Culpeper County, two localities), Orange (Madison County), Falls Church, 6 miles west of Lawrenceville (Brunswick County), Norwood, Amelia Court House (two localities, Amelia County), Caledonia (Goochland County), and Clifton (Fairfax County), Va.; Fawn Knob, Yancey County, N. C.; Roane Mountain, Tenn.; Coon Creek, Tallapoosa County, and Jefferson County, Ala.; Clam River, Burnett County, Wis.; Jacumba (San Diego County) and Santa Catalina Island, Calif.; Columbia River above Kettle Falls, Wash.; Pipestone Creek, southwest Montana; bordering Buena Vista Hills, San Joaquin Valley, Calif.; Tuolumne, Calif.; 4 miles northeast of Lindsay, Calif.

So,   Sodalite:

Cerro Sapo, Cochabamba Cordillera, Bolivia.

Su,   Sunstone:

Reservation near Globe, Ariz.

To,   Tourmaline:

Mesa Grande, San Diego County, Calif.

T,   Turquoise:

Los Cerrillos, Burro Mountains, Hachita Mountains, Jarilla Mountains, and Paschal, N. Mex.; Sugar Loaf Peak (Lincoln County), Columbus (Mineral County), Crescent (Clark County), Royston (Nye County), near Boulder City, and northeast of Searchlight, Landon County, Nev.; Turquoise Mountain, Cochise County, Ariz.; Mineral Park, Mohave County, Ariz.; Manvel, also Silver Lake, San Bernardino County, Calif.; La Jara, Conejos County, Colo.; Chuquicamata, Chile.

BIBLIOGRAPHY

ACOSTA, JOSEPH DE.

1880. The natural and moral history of the Indies. Edited by Clements R. Markham. 2 vols. Hakluyt Soc. Publ. [Nos. 60–61.] London.

ADAMS, JOHN.

1813. A voyage to South America. [Ulloa's voyage.] Pinkerton's Voyages, vol. 14, pp. 313–696. London.

AHLFELD, F., and MOSEBACH, R.

1935. Neues Jahrb. f. Min., Geol. u. Paläontol., Beil-Bd. 69, Abt. A (3), pp. 388–404. Heidelberg and Stuttgart.

AHLFELD, FR., and WEGNER, R. N.

1932. Über die Herkunst der im Bereich altperuanischer Kulturen gefundenen Schmuckstücke aus Sodalith. Zeitschr. f. Ethnol., vol. 63, pp. 288–296. Berlin.

ALARCHON, HERNANDO DE.

1904. The relation of the navigation and discovery which Captaine Fernando Alarchon made by order of . . . Don Antonio de Mendoça, Vizeroy of New Spaine [1540]. Hakluyt's Voyages, vol. 9, pp. 279–318 Glasgow.

ALTER, J. CECIL.

1925. James Bridger. Salt Lake City.

Anonymous.
1891. Mexican onyx mines. Eng. and Min. Journ., vol. 52, No. 26, p. 729.
1897. Turquoise in Nevada. *Paragraph in* Eng. and Min. Journ., vol. 64, No. 16, p. 456.
1902. Turquoise mining in Arizona and New Mexico. Min. and Sci. Press, vol. 85, No. 8, pp. 102–103. San Francisco.
1937. The Aztec eagle. Jewelers' Circular-Keystone, vol. 108, No. 1, pp. 97, 99, 101, Oct.

Armstrong, Moses K.
1901. The early empire builders of the Great West. St. Paul.

Ayer, E. E., Translator.
1916. The memorial of Fray Alonso de Benavides, 1630. Chicago.

Ball, Sydney H.
1931. Historical notes on gem mining. Econ. Geol., vol. 26, pp. 681–738.

[Bancroft, Edward.]
1769. Essay on the natural history of Guiana . . . interspersed with a variety of literary and medical observations in several letters from a gentleman of the Medical Faculty. London.

Bancroft, Herbert Howe.
1886. The works of. 39 vols., 1886–1890. The native races, vols. 1–5. San Francisco.

Bandelier, A. F.
1890. Contributions to the history of the southwestern portion of the United States. Pap. Archeol. Inst. Amer., Amer. Ser., vol. 5.
1890 a. The delight makers. New York.

Barber, Edwin A.
1883. Catlinite, its antiquity as a material for tobacco pipes. Amer. Nat., vol. 17, pp. 745–764.

Barrett, S. A.
1926. Field studies for the catlinite and quartzite groups. Yearbook Pub. Mus. City of Milwaukee, 1924, pp. 7–21.
1933. Pomo myths. Bull. Pub. Mus. City of Milwaukee, vol. 15.

Bartlett, J. C.
1854. Personal narrative of explorations and incidents . . . connected with the United States and Mexican Boundary Commission. 4 vols. New York.

Bauer, Max H.
1900. Fuchsit als Material zu prähistorischen Artefacten aus Guatemala. Centr. f. Min. Geol. u. Palaeon., pp. 291–292. Stuttgart.
1904. Precious stones. Trans. fr. German by L. J. Spencer. London.

Benavides, Alonso de.
1916. The memorial of Fray Alonso de Benavides, 1630. Trans. by Mrs. Edward E. Ayer. Chicago.

Benedict, Ruth.
1931. Tales of the Cochiti Indians. Bur. Amer. Ethnol. Bull. 98.

Bennett, Wendell C., and Zingg, Robert M.
1935. The Tarahumara, an Indian tribe of northern Mexico. Chicago.

Benzoni, Girolamo.
1857. History of the New World. Trans. and edited by W. H. Smyth. Hakluyt Soc. Publ. [No. 21], London.

Berton, Adrien.
1936. The Corocoro copper district of Bolivia. Amer. Inst. Min. and Met. Eng., Tech. Publ. No. 698, New York.

Beuchat, H.
　　1912. Manuel d'archéologie américaine. Paris.
Biart, Lucien.
　　1887. The Aztecs; their history, manners, and customs. Trans. by J. L. Garner. Chicago.
Blake, W. P.
　　1858. The chalchihuitl of the ancient Mexicans: its locality and association and its identity with turquois. Amer. Journ. Sci., 2d ser., vol. 25, pp. 227–232.
　　1883. New locality of the green turquois known as chalchuite. . . . Amer. Journ. Sci., 3d ser., vol. 25, pp. 197–200.
Blatchley, W. S.
　　1897. Indiana caves and their fauna. 21st Ann. Rep., Indiana Dept. Geol. and Nat. Res., pp. 121–212.
Blondel, S.
　　1876. Cartas de Hernando Cortes. Rev. Phil. et Ethnogr., T. 2, p. 294. Paris.
Boas, Franz.
　　1907. The Eskimo of Baffin Land and Hudson Bay. Bull. Amer. Mus. Nat. Hist., vol. 15.
Boddam-Whetham, J. W.
　　1877. Across Central America. London.
Bollaert, William.
　　1860. Antiquarian, ethnological, and other researches in New Granada, Ecuador, Peru, and Chili, with observations on the pre-Incarial, Incarial, and other monuments of the Peruvian nations. London.
Bolton, Herbert E., Editor.
　　1914. Athanase de Mézières and the Louisiana-Texas frontier, 1768–1780. 2 vols., Cleveland.
　　1916. Spanish exploration in the Southwest, 1542–1706. New York.
Bourke, John G.
　　1890. Vesper hours of the Stone Age. Amer. Anthrop., vol. 3, pp. 55–63.
　　1892. The medicine-men of the Apache. 9th Ann. Rep. Bur. Amer. Ethnol., 1887–88, pp. 443–603.
Bradbury, Jno.
　　1904. Travels in the interior of America, 1809–1811. Early western travels, 1748–1846, vol. 5. Edited by Reuben Gold Thwaites. Cleveland.
Brendler, Wolfgang.
　　1934. Sodalite from Bolivia. Amer. Mineralogist, vol. 19, pp. 28–31.
Breton, Adela.
　　1905. Some obsidian workings in Mexico. Proc. Int. Congr. Amer., 13th sess., New York, 1902, pp. 265–268.
Brigham, William T.
　　1887. Guatemala: the land of the quetzál. New York.
Brinton, Daniel G.
　　1883. The folk-lore of Yucatan. Folk-Lore Journ., vol. 1, pp. 244–256.
Brower, J. V.
　　1897. The Missouri River and its utmost source. St. Paul.
Brown, Charles E.
　　1914. A Wisconsin catlinite quarry. Wis. Archeologist, vol. 13, pp. 80–82.
Brown, Charles E., and Skavlem, H. L.
　　1914. Notes on some archeological features of Eau Claire, Cheppewa, Rusk, and Dunn Counties. Wis. Archeologist, vol. 13, pp. 60–79.
Browne, P. W.
　　1909. Where the fishers go; the story of Labrador. New York.

Bunzel, Ruth L.
    1932. Zuñi Katcinas. 47th Ann. Rep. Bur. Amer. Ethnol., 1929–30, pp. 837–1086.
Burpee, Lawrence J.
    1908. The search for the Western Sea; the story of the exploration of north-western America. Toronto.
Burton, Richard Francis.
    1869. The highlands of the Brazil. 2 vols. London.
Camsell, Charles.
    1912. The mineral resources of a part of the Yale District, B. C. Journ. Canadian Min. Inst., 1911, vol. 14, pp. 596–611. Montreal.
Cartwright, Captain. See Townsend, C. W., Editor.
Carver, Jonathan.
    1778. Travels through the interior parts of North America in the years 1766, 1767, and 1768. London.
Caso, Alfonso.
    1932. Monte Albán, richest archeological find in America, Nat. Geogr. Mag., vol. 62, pp. 487–512, Oct.
Catlin, George.
    1848. Illustrations of the manners, customs, and condition of the North American Indians: in a series of letters and notes written during eight years of travel . . . among the wildest . . . tribes. 2 vols. 7th ed., London.
    1913. Illustrations of the manners and customs of the North American Indians. 2 vols. London.
Chambers, Ernest J.
    1914. The unexploited West. Ottawa.
Chapin, Frederick H.
    1892. The land of the Cliff Dwellers. Boston. 1892.
Charlevoix, Pierre F. X. de.
    1763. Letters to the Dutchess of Lesdiguieres, giving an account of a voyage to Canada and travels through that country and Louisiana to the Gulf of Mexico. London.
Charnay, Désiré.
    1887. The ancient cities of the New World. New York.
Chase, Henry E.
    1885. Notes on the Wampanoag Indians. Ann. Rep. Smithsonian Inst. 1883, pp. 878–907.
Church, George Earl.
    1912. Aborigines of South America. Edited by Clements R. Markham. London.
Clark, W. M.
    1878. Antiquities of Tennessee. Ann. Rep. Smithsonian Inst. 1877, pp. 269–276.
Clarke, F. W., and Chatard, T. M.
    1884. Mineral notes from the laboratory of the U. S. Geological Survey. Amer. Journ. Sci., 3d ser., vol. 28, pp. 20–25.
Clarke, F. W., and Merrill, G. P.
    1888. On nephrite and jadeite. Proc. U. S. Nat. Mus., vol. 11, pp. 115–130.
Clavigero, Abbé d'Francisco S.
    1807. The history of Mexico. Trans. by Charles Cullen. London.
Cogolludo, D. L. de.
    1688. Historia de Yucathan. Madrid.

Columbia Emerald Syndicate, Ltd., Pamphlet of.
   1921. New York City.
Coolidge, Mary Roberts.
   1929. The rain-makers.   Indians of Arizona and New Mexico.   Boston and
      New York.
Cooper, John M.
   1917. Analytical and critical bibliography of the tribes of Tierra del Fuego
      and adjacent territory.   Bur. Amer. Ethnol. Bull. 63.
Corbusier, William F.
   1886. The Apache-Yumas and Apache Mojaves.   Amer. Antiq. and Orient.
      Journ., vol. 8, pp. 276–284.   September.
Cosgrove, H. S., and C. B.
   1932. The Swarts ruin.   Pap. Peabody Mus. Amer. Archaeol. and Ethnol.,
      vol. 15, No. 1.
Coues, Elliott, Editor.
   1893. History of the expedition of Lewis and Clarke to the sources of the
      Missouri River and to the Pacific in 1804–5–6.   A new ed.   4 vols.
      New York.
   1895. The expeditions of Zebulon Montgomery Pike.   3 vols.   New York.
Coxe, William.
   1787. Account of the discoveries between Asia and America.   3d ed.   London.
Crawford, William P., and Johnson, Frank.
   1937. Turquoise deposits of Courtland, Arizona.   Econ. Geol., vol 32,
      pp. 511–523.
Curtis, Natalie.
   1907. The Indians' book.   New York.
Cushing, F. H.
   1901. Zuñi folk-tales.   New York.
Dall, William H.
   1870. Alaska and its resources.   Boston.
Davis, Emily C.
   1931. Ancient Americans; the archaeological story of two continents.   New
      York.
Dawson, George M.
   1887. Note on the occurrence of jade in British Columbia and its employment
      by the natives.   Can. Rec. Sci., vol. 2, No. 6.   April.
   1888. Geological observations of the Yukon Expedition, 1887.   Science,
      vol. 9, pp. 185–186.
Deisher, Henry.
   1932. South Mountain Indian quarries.   Bucks County [Pa.] Hist. Soc.,
      Coll. of Papers, vol. 6, pp. 334–341.   Allentown. Pa.
De la Ronde, John T.
   1876. Personal narrative.   Coll. State Hist. Soc. Wis., vol. 7, pp. 345–365.
De Nadaillac, Marquis.
   1884. Pre-historic America.   Trans. by N. D'Anvers.   New York.
Denison, Fred.   See Chase, Henry E.
Densmore, Frances.
   1929. Chippewa customs.   Bur. Amer. Ethnol. Bull. 86.
De Smet, P. J.
   1847. Oregon Missions and travels over the Rocky Mountains, 1845–46.
      New York.
Dodge, R. I.
   1877. The plains of the Great West and their inhabitants.   With an intro-
      duction by William Blackmore.   New York.

DOMENECH, EMMANUEL.
 1860. Seven years residence in the Great Deserts of North America. 2 vols. London.
DONOHOE, THOMAS.
 1895. The Iroquois and the Jesuits. Buffalo.
DORSEY, GEORGE A.
 1901. Archaeological investigations on the Island of La Plata, Ecuador. Field Columbian Mus., Publ. 56, Anthrop. Ser., vol. 2, No. 5. April.
DYOTT, G. M.
 1923. Silent highways of the jungle: being the record of an adventurous journey across Peru to the Amazon. London.
ELEVENTH CENSUS, 1890, P. 186.
EMMONS, GEORGE T.
 1923. Jade in British Columbia and Alaska and its use by the natives. Ind. Notes and Monogr., No. 35, Mus. Amer. Ind., Heye Foundation.
FARABEE, WILLIAM CURTIS.
 1917. A pioneer in Amazonia: the narrative of a journey from Manaoes to Georgetown. Bull. Geogr. Soc. Phila., vol. 15, No. 2.
FARRINGTON, O. C.
 1903. Gems and gem minerals. Chicago.
FEWKES, JESSE WALTER.
 1898. Archeological expedition to Arizona in 1895. 17th Ann. Rep. Bur. Amer. Ethnol., 1895–96, pt. 2, pp. 519–744.
 1904. Two summers' work in Pueblo ruins. 22d Ann. Rep. Bur. Amer. Ethnol., 1900–1901, pt. 1, pp. 3–195.
FICKLEN, J. R. See Fortier, Alcée, and Ficklen, J. R.
FITZROY, ROBERT.
 1839. Narrative of the surveying voyages of His Majesty's ships Adventure and Beagle between the years 1826 and 1836. 3 vols. London.
FONTAINE, WM. F.
 1883. Notes on the occurrence of certain minerals in Amelia County, Virginia. Amer. Journ. Sci., 3d ser., vol. 25, pp. 330–339.
FORBES, P. L.
 1935. Iridescent obsidian. Gemmologist, vol. 4, No. 46, pp. 306–309. May.
FORDE, C. DARYLL.
 1831. Ethnography of the Yuma Indians. Univ. Calif. Publ. Amer. Archaeol. and Ethnol., vol. 28, No. 4, pp. 83–278. Dec. 12.
FORTIER, ALCÉE, and FICKLEN, J. R.
 1907. Central America and Mexico. The history of North America, vol. 9. Philadelphia.
FOSTER, J. W.
 1873. Pre-historic races of the United States of America. Chicago.
FOWKE, GERARD.
 1888–89. Some popular errors in regard to Mound Builders and Indians. Ohio Archaeol. and Hist. Quart., vol. 2, pp. 380–403.
 1888–89 a. The manufacture and use of aboriginal stone implements. Ohio Archaeol. and Hist. Quart., vol. 2, pp. 514–533.
 1894. Archeologic investigations in James and Potomac Valleys. Bur. Amer. Ethnol. Bull. 23.
 1922. Archeological investigations. Bur. Amer. Ethnol. Bull. 76.
FRANKLIN, JNO.
 1823. Narrative of a journey to the shores of the Polar Sea, in the years 1819–20–21–22. London.

FRENCH, B. F.
  1846–1853. Historical collections of Louisiana, pts. 1–5. New York.
  1875. Historical collections of Louisiana and Florida, 2d ser., Memoirs and
    narratives. New York.
FULTON, A. R.
  1882. The Red Men of Iowa. Des Moines.
GANN, THOMAS.
  1925. Maya jades. Proc. 21st Int. Congr. Amer. Göteborg 1924, pt. 2, pp.
    274–282. Göteborg.
  1926. Ancient cities and modern tribes: exploration and adventure in Maya
    lands. New York.
GARCIA, ICAZBAL-CETA JOAQUIN.
  1936. Soc. Mex. Geogr., Bol. 2, da epoca t. 4, p. 559.
GARCILASO (GARCILLASSO) DE LA VEGA.
  1688. The royal commentaries of Peru. Trans. by Sir Paul Rycaut. Pts.
    1–2 [in 1 vol.]. London.
  1869. Royal commentaries of the Yncas, vol. 1. Hakluyt Soc. Publ. [No. 41].
    London.
  1871. Royal commentaries of the Yncas, vol. 2. Hakluyt Soc. Publ. [No. 45].
    London.
GIFFORD, E. W.
  1931. The Kamia of Imperial Valley. Bur. Amer. Ethnol. Bull. 97.
  1932. The southeastern Yavapai. Univ. Calif. Publ. Amer. Archaeol. and
    Ethnol., vol. 29, No. 3, pp. 176–252, Feb. 6.
GILLIN, JOHN.
  1936. The Barama River Caribs of British Guiana. Pap. Peabody Mus.
    Amer. Archaeol. and Ethnol., vol. 14, No. 2.
GILMORE, MELVIN RANDOLPH.
  1929. Prairie smoke. New York.
GODDARD, PLINY EARLE.
  1927. Indians of the Southwest. New York.
GOLDENWEISER, ALEXANDER A.
  1922. Early civilization; an introduction to anthropology. New York.
GRAVIER, JACQUES.
  1900. Relation or journal of the voyage of Father Gravier of the Society of
    Jesus, in 1700, from the country of the Illinois to the mouth of the
    Mississippi River. In Jesuit relations and allied documents, vol. 65,
    pp. 100–179. (See Thwaites.)
GRIEG, SIGURD.
  1932. Aarbok, 1930. Universitet, Oslo, Norway.
HAGIE, C. E.
  1936. Interglacial man in America. Sci. Amer., vol. 154, No. 6, p. 325, June.
HAKLUYT, RICHARD, COMPILER.
  1850. Divers voyages touching the discovery of America and the islands
    adjacent. Collected and published by Richard Hakluyt in the year
    1582. Edited by John Winter Jones. Hakluyt Soc. Publ. [No. 7].
    London.
  1903–1905. The principal navigations, voyages, traffiques, and discoveries of
    of the English Nation. 12 vols. Glasgow.
HARRINGTON, JOHN PEABODY.
  1916. The ethnogeography of the Tewa Indians. 29th Ann. Rep. Bur. Amer.
    Ethnol., 1907–8, pp. 29–636.

HARRINGTON, M. R.
    1913. Preliminary sketch of Lenape culture. Amer. Anthrop., n. s., vol. 15,
        pp. 208–235.
    1924. A West Indian gem center. Ind. Notes, vol. 1, No. 4, pp. 184–189,
        Mus. Amer. Ind., Heye Foundation.
    1930. Archeological explorations in southern Nevada. Rep. 1st. Sessions
        Exped., 1929. Southwest Mus. Pap., No. 4, June.
HARTMAN, C. V.
    1901. Archaeological researches in Costa Rica. Roy. Ethnol. Mus. Stock-
        holm.
    1907. Archaeological researches on the Pacific coast of Costa Rica. Mem.
        Carnegie Mus., vol. 3. Pittsburgh.
HAYDEN, F. V.
    1867. Sketch of the geology of northeastern Dakota, with a notice of a short
        visit to the celebrated Pipestone quarry. Amer. Journ. Sci., 2d
        ser., vol. 43, pp. 15–22.
HEGER, FRANZ.
    1925. Klanzplatten von Nephrit aus Venezuela. Proc. 21st Congr. Amer.
        Göteborg 1924, pt. 2, pp. 148–155. Göteborg.
HIDDEN, W. E.
    1893. Two new localities for turquoise. Amer. Journ. Sci., 3d ser., vol. 46,
        pp. 400–402.
HISTORY OF JO DAVIESS COUNTY, ILLINOIS.
    1878. [Published by H. F. Kett & Co.] Chicago.
HODGE, F. W.
    1922. Guide to the Museum of the American Indian, Heye Foundation,
        Second floor. Ind. Notes and Monogr., Mus. Amer. Ind., Heye
        Foundation.
HOLMES, W. H.
    1892. Sacred pipestone quarries of Minnesota and ancient copper mines of
        Lake Superior. Proc. Amer. Assoc. Adv. Sci., vol. 41, pp. 277–279.
    1897. Stone implements of the Potomac-Chesapeake tidewater province.
        15th Ann. Rep. Bur. Amer. Ethnol., pp. 3–152.
    1900. The obsidian mines of Hidalgo, Mexico. Amer. Anthrop., n. s., vol. 2,
        pp. 405–416.
    1904. Traces of aboriginal operations in an iron mine near Leslie, Mo. Ann.
        Rep. Smithsonian Inst. 1903, pp. 723–726.
    1906. Certain notched or scalloped stone tablets of the Mound-Builders.
        Amer. Anthrop., n. s., vol. 8, pp. 101–108.
    1907. [Article] Amber, in Handbook of American Indians north of Mexico.
        Bur. Amer. Ethnol. Bull. 30, pt. 1, p. 48.
    1919. Handbook of aboriginal American antiquities. Pt. 1. Introductory.
        The lithic industries. Bur. Amer. Ethnol. Bull. 60.
HOUGH, WALTER.
    1890. Fire-making apparatus in the U. S. National Museum. Rep. U. S.
        Nat. Mus. 1888, pp. 531–587.
    1902. A collection of Hopi ceremonial pigments. Rep. U. S. Nat. Mus. 1900,
        pp. 465–471.
    1903. Archeological field work in northeastern Arizona. The Museum-
        Gates Expedition of 1901. Rep. U. S. Nat. Mus. 1901, pp. 279–358.
    1907. Antiquities of the upper Gila and Salt River Valleys in Arizona and
        New Mexico. Bur. Amer. Ethnol. Bull. 35.
HOWARD, J. H.
    1936. The ancient lapidary. Rocks and minerals. p. 78. May.

Humboldt, Friederich H. Alexander de.

 1811. Essai politique sur le royaume de la Nouvelle-Espagne. 2d ed., 6 vols. Paris.

 1814–29. Personal narrative of travel to the equinoctial regions of the new continent during the years 1799–1804. Trans. by H. M. Williams. 7 vols. London.

 1815. Vues des Cordillères et monumens des peuples indigènes de l'Amérique. 2 vols. Paris.

Im Thurn, E. F.

 1883. Among the Indians of Guiana. London.

Irving, John T.

 1888. Indian sketches taken during an expedition to . . . the Pawnee and other tribes in 1833. First published in 1835. 2 vols. New York.

James, George Wharton.

 1903. The Indians of the Painted Desert region. Boston.

 1920. New Mexico, the land of the Delight Makers. Boston.

Jeancon, J. A.

 1923. Excavations in the Chama Valley, New Mexico. Bur. Amer. Ethnol. Bull. 81.

Jenness, D.

 1922. The life of the Copper Eskimo. Ottawa.

 1925. A new Eskimo culture in Hudson Bay. Geogr. Rev., vol. 15, No. 3, pp. 428–437.

 1932. The Indians of Canada. Nat. Mus. Canada, Bull. 65, Anthrop. Ser., No. 15.

Johnson, Douglas Wilson.

 1903. The geology of the Cerrillos Hills, New Mexico. Pt. 1, General geology. (Cont.) School Mines Quart., vol. 24, No. 4, pp. 456–500.

 1904. The geology of the Cerrillos Hills, New Mexico. Pt. 3, Petrography. School Mines Quart., vol. 25, pp. 69–98.

Jones, C. C.

 1859. Indian remains in southern Georgia. *Address before* Ga. Hist. Soc., Feb. 12, 1859, 25 pp. Savannah.

Jourdanet, D. *See* Sahagun.

Joyce, Thomas A.

 1912. South American archaeology; an introduction to the archaeology of the South American continent with special reference to the early history of Peru. London.

 1914. Mexican archaeology; an introduction to the archaeology of the Mexican and Mayan civilizations of pre-Spanish America. New York.

 1916. Central American and West Indian archaeology; being an introduction to the archaeology of the states of Nicaragua, Costa Rica, Panama, and the West Indies. London.

Judd, Neil M.

 1925. Everyday life in Pueblo Bonito. Nat. Geogr. Mag., vol. 48, No. 3, pp. 227–262. September.

Kalm, Peter.

 1772. Travels into North America. 2 ed., 2 vols. London.

Keating, W. H.

 1824. Narrative of an expedition to the source of St. Peter's River, compiled from the notes of Major H. Long [and others]. 2 vols. Philadelphia.

KERBEY, JOSEPH ORTON.
   1906. The land of tomorrow. New York.
KERR, ROBERT.
   1811. A general history and collections of voyages and travels. [18 vols.]
      Vol. 3. Edinburgh.
KERR, W. C.
   1875. Report of the Geological Society of North Carolina, vol. 1. Raleigh.
   1880. The mica veins of North Carolina. Trans. Amer. Inst. Min. Eng.,
      vol. 8, pp. 457–462.
KEYMIS, LAURENCE.
   1904. A relation of the second voyage to Guiana performed and written in
      the yeere 1596. Hakluyt's Voyages, vol. 10, pp. 441–501. Glas-
      gow.
KIDDER, ALFRED VINCENT.
   1932. The artifacts of Pecos. Peabody Found. Archaeol., Phillips Acad.,
      Andover, Mass.
KIDDER, A. V., and GUERNEY, S. J.
   1919. Archeological explorations in northeastern Arizona. Bur. Amer.
      Ethnol. Bull. 65.
KING, FRANCIS P.
   1894. A preliminary report on the corundum deposits of Georgia. Geol.
      Surv. Georgia, Bull. No. 2.
KNIFFEN, FRED B.
   1928. Achomawi geography. Univ. Calif. Publ. Amer. Archaeol. and
      Ethnol., vol. 23, No. 5, pp. 297–332, Jan. 14.
KNIGHT, WILBUR C.
   1898. Prehistoric quartzite quarries in central eastern Wyoming. Science,
      n. s., vol. 7, pp. 308–311.
KOTZEBUE, OTTO VON.
   1821. Voyage of discovery into the South Sea and Behrings Straits in 1815–18.
      3 vols. London.
KROEBER, A. L.
   1925. Handbook of the Indians of California. Bur. Amer. Ethnol. Bull. 78.
   1931. The Seri. Southwest Mus. Pap., No. 6, April.
KUNZ, GEORGE FREDERICK.
   1890. Gems and precious stones of North America. New York.
   1899. Precious stones. 20th Ann. Rep. U. S. Geol. Surv., 1898–99. Pt. 6
      (cont.), pp. 557–600.
   1902. Precious stones. U. S. Geol. Surv., Min. Res. U. S., 1901. Pt. 2,
      Nonmetallic products, pp. 729–771
   1905. Semiprecious stones, gems, jewelers' materials, and ornamental stones
      of California. 2d ed. Sacramento.
   1907. History of gems found in North Carolina. N. C. Geol. and Econ.
      Surv. Bull. No. 12.
KURZ, RUDOLPH FRIEDERICH.
   1937. Journal of Rudolph Friederich Kurz. Trans. by Myrtis Jarrell, edited
      by J. N. B. Hewitt. Bur. Amer. Ethnol. Bull. 115.
LAIDLAW, GEORGE E.
   1897. Balsam Lake. Archeol. Rep. 1896–97. App. Rep. Min. Education,
      Ontario, Can., pp. 80–89, Toronto.
   1897 a. The aboriginal remains of Balsam Lake, Ontario. Amer. Antiq.,
      vol. 19, No. 2, pp. 68–72.
LANMAN, CHARLES.
   1856. Adventures in the wilds of the United States and British American
      provinces. 2 vols. Philadelphia.

Latcham, R. E.
　　1909. Ethnology of the Araucanos. Journ. Roy. Anthrop. Inst., vol. 39,
　　　　pp. 334–370.
Laudonniere, René.
　　1904. A notable historie containing foure voyages . . . into Florida.
　　　　In Hakluyt, Richard, Voyages, vol. 8, pp. 439–486. London.
Laut, Agnes C.
　　1918. Pathfinders of the West. New York.
La Vérendrye, Pierre G. de.
　　1927. Journals and letters of Pierre G. de La Vérendrye and his sons.
　　　　Edited by L. J. Burpee. Publ. Champlain Soc. Toronto.
Leffingwell, Ernest De K.
　　1919. The Canning River region, northern Alaska. U. S. Geol. Surv., Prof.
　　　　Pap. 109.
Le Page du Pratz, Antoine S.
　　1758. Histoire de la Louisiane. T. 1–3, Paris.
LeRoy, Claude Charles.
　　1753. History of the savage peoples who are allies of New France. Paris.
Lilly, Eli.
　　1937. Prehistoric antiquities of Indiana. Indianapolis.
Lincoln, F. C.
　　1923. Mining districts and mineral resources of Nevada.
　　　　Reno.
Lindgren, Waldemar; Graton, Louis C.; and Gordon, Charles.
　　1910. The ore deposits of New Mexico. U. S. Geol. Surv., Prof. Pap. 68.
Loeb, Edwin M.
　　1926. Pomo folkways. Univ. Calif. Publ. Amer. Archaeol. and Ethnol., vol.
　　　　19, No. 2, pp. 149–405, Sept. 29.
Loskiel, George Henry.
　　1789. Geschichte der Mission der Evangelischen Brüder unter den Indianern
　　　　in Nordamerika. Barby.
Lothrop, Samuel Kirkland.
　　1928. The Indians of Tierra del Fuego. Contr. Mus. Amer. Ind., Heye
　　　　Foundation, vol. 10.
　　1937. Coclé, an archaeological study of Central Panama. Mem. Peabody
　　　　Mus. Amer. Archaeol. and Ethnol., vol. 7. pt. 1.
Löwenstern, Isidore.
　　1843. Le Mexique. Paris.
Lummis, Charles F.
　　1892. Some strange corners of our country. New York.
　　1925. Mesa, cañon, and pueblo; our wonderland of the Southwest. New
　　　　York.
Lynd, James W.
　　1865. History of the Dakotas. Minn. Hist. Soc. Coll., vol. 2, pp. 57–84.
　　　　St. Paul.
Lyon, G. F.
　　1825. A brief narrative of an unsuccessful attempt to reach Repulse Bay.
　　　　London.
MacCreagh, Gordon.
　　1926. White waters and black. New York.
McGovern, William Montgomery.
　　1927. Jungle paths and Inca ruins. New York.

MCKENNEY, THOMAS L., and HALL, JAMES.
  1833. The Indian tribes of North America. Edited by F. W. Hodge. Edinburgh.
MACKENZIE, ALEXANDER.
  1902. Voyages from Montreal through the continent . . . to the frozen and Pacific Oceans in 1789 and 1793. 2 vols. New York.
MALLERY, GARRICK.
  1893. Picture writing of the American Indians. 10th Ann. Rep. Bur. Amer. Ethnol., 1888–89, pp. 3–807.
MARKHAM, C. R. See Garcilaso de la Vega; Col. George E. Church.
MASON, GREGORY.
  1926. Cities that passed in a night. New light on the tragic history of the Mayas. The World's Work, vol. 52, No. 4, pp. 433–439, August.
  1931. Columbus came late. New York.
MASON, J. ALDEN.
  1936. Archaeology of Santa Marta. The Tairona culture. Pt. 2, sec. 1. Objects of stone, shell, bone, and metal. Field Mus. Nat. Hist., Publ. 358, Anthrop Ser., vol. 20, No. 2.
MATHIASSEN, THERKEL.
  1927. Archaeology of the Central Eskimos. Copenhagen.
  1927 a. Archaeology of the Central Eskimos. Pt. 2. The Thule culture and its position within the Eskimo culture. Copenhagen.
MATTHEWS, WASHINGTON.
  1877. Ethnography and philology of the Hidatsa Indians. U. S. Geol. and Geogr. Surv. Misc. Publ. No. 7.
MAXIMILIAN, ALEX. P., PRINCE OF WEID.
  1843. Travels in the interior of North America. London.
MEANS, PHILIP AINSWORTH.
  1931. Ancient civilizations of the Andes. New York.
MEEKER, MOSES.
  1872. Early history of the lead region of Wisconsin. Coll. State Hist. Soc. Wis., vol. 6, pp. 271–296. Madison.
MENDEZ, SANTIAGO. See SAVILLE, M. H., EDITOR.
MERCER, H. C.
  1894. Indian jasper mines in the Lehigh hills. Amer. Anthrop., vol. 7, pp. 80–92.
  1895. Jasper and stalagmite quarried by Indians in the Wyandotte Cave. Proc. Amer. Phil. Soc., vol. 34, pp. 396–400.
MITCHENER, C. H.
  1876. Ohio Annals. Historical events in the Tuscarawas and Muskingum Valleys and in other portions of the State of Ohio. Dayton, Ohio.
MOLINA, ALONSO DE
  1571. Vocabulario en lengua Castellana y Mexicana. Mexico.
MOLINA, JUAN IGNACIO.
  1809. The geographical, natural, and civil history of Chile. London.
MÖLLHAUSEN, BALDWIN.
  1858. Dairy of a journey from the Mississippi to the coasts of the Pacific. 2 vols. London.
MONTESINOS, FERNANDO.
  1920. Memorias antiguas historiales del Peru. Trans. by Philip Ainsworth Means. Hakluyt Soc. Publ., 2d ser., No. 48. London.

MOOREHEAD, WARREN K.
1910. The Stone Age in North America. 2 vols. Boston.
1917. Stone ornaments used by the Indians in the United States and Canada. Andover, Mass.

MORGAN, LEWIS HENRY.
1901. League of the Ho-dé-no-sau-nee or Iroquois. Edited by Herbert M. Lloyd. New York.

MORLEY, MARGARET W.
1913. The Carolina Mountains. Boston.

MORRIS, EARL H.
1919. Preliminary account of the antiquities of the region between the Mancos and La Plata Rivers in southwestern Colorado. 33d Ann. Rep. Bur. Amer. Ethnol., 1911–12, pp. 155–206.
1929. An aboriginal salt mine at Camp Verde, Arizona. Anthrop. Pap. Amer. Mus. Nat. Hist., vol. 30, pp. 81–97.
1931. The Temple of the Warriors; the adventure of exploring . . . the ruined city of Chichen Itza, Yucatan. New York.

MUSTERS, GEORGE C.
1871. At home with the Patagonians. London.

MURDOCH, JOHN.
1892. Ethnological results of the Point Barrow Expedition. 9th Ann. Rep. Bur. Amer. Ethnol., 1887–88, pp. 3–441.

MYERS, HENRY M., and MYERS, PHILIP VAN NESS.
1871. Life and nature under the tropics. New York.

NEILL, EDWARD DUFFIELD.
1858. The history of Minnesota from the earliest French explorations to the present time. Philadelphia.

NELSON, EDWARD WILLIAM.
1899. The Eskimo about Bering Strait. 18th Ann. Rep. Bur. Amer. Ethnol., 1896–97, pt. 1, pp. 3–518.

NEWBERRY, JOHN S.
1876. Geological Report. *In* Report of the Exploring expedition from Santa Fe, New Mexico, to the junction of the Grand and Green Rivers . . . in 1859, under the command of Captain J. N. Macomb. Washington.
1892. The ancient civilizations of America. Pop. Sci. Monthly, vol. 41. pp. 187–200.

NICOLLET, L. N.
1843. Report intended to illustrate map of the hydrographic basin of the upper Mississippi River. 26th Congr., 2d Sess., S. Doc. 237.

NUTTALL, ZELIA.
1901. Chalchihuitl in ancient Mexico. Amer. Anthrop., n. s., vol. 3, pp. 227–238.
1901 a. The fundamental principles of Old and New World civilizations. Pap. Peabody Mus. Amer. Archaeol. and Ethnol., vol. 2.

OLBRECHTS, FRANS M.
1930. Some Cherokee methods of divination. Proc. 23d Int. Congr. Amer. New York, 1928, pp. 547–552.

OLDEN, CHARLES.
1912. Emeralds: Their mode of occurrence and methods of mining and extraction in Colombia. Trans. Inst. Min. and Met., vol. 21, pp. 193–203.

PACKARD, A. S.
    1891. The Labrador coast. A journal of two summer cruises to that region.
          New York.
PACKARD, R. L.
    1893. Pre-Columbian mining in North America. Amer. Antiq. and Orient.
          Journ., vol. 15, pp. 152–164, May.
PARSONS, ELSIE CLEWS.
    1928. Notes on the Pima, 1926. Amer. Anthrop., n. s., vol. 30, pp. 445–464.
    1932. Isleta, New Mexico. 47th Ann. Rep. Bur. Amer. Ethnol., 1929–30,
          pp. 193–466.
PEABODY, CHARLES.
    1904. Exploration of mounds, Coahoma County, Mississippi. Pap. Pea-
          body Mus. Amer. Archaeol. and Ethnol., vol. 3, No. 2, pp. 23–63.
PEÑAFIEL, ANTONIO.
    1890. Monuments of ancient Mexican art. Berlin.
PEPPER, GEORGE H.
    1905. Ceremonial objects and ornaments from Pueblo Bonita, New Mexico.
          Amer. Anthrop., n. s., vol. 7, pp. 183–197.
    1909. The exploration of a burial-room in Pueblo Bonito, New Mexico.
          Putnam Ann. Vol., pp. 196–252.
    1920. Pueblo Bonito. Anthrop. Pap. Amer. Mus. Nat. Hist., vol. 27.
PERROT, NICOLAS.
    1911. Memoir on the manners, customs, and religion of the savages of North
          America. In The Indian tribes of the upper Mississippi Valley
          and region of the Great Lakes. Vol. 1, pp. 25–272. Trans. by
          Emma Helen Blair. Cleveland.
PETROFF, IVAN.
    1884. Report on the population, industries, and resources of Alaska. U. S.
          Dept. Interior, Census Office, 10th Census, vol. 8.
PHILLIPS, WILLIAM B.
    1888. Mica mining in North Carolina. U. S. Geol. Surv., Min. res. U. S.,
          1887, pp. 661–671.
PICKETT, A. J.
    1851. History of Alabama, and incidentally of Georgia and Mississippi from
          the earliest period. 3d ed., 2 vols., Charleston.
PIKE, ZEBULON M. See COUES, ELLIOTT, EDITOR.
PINKERTON, JOHN.
    1808–14. General collection of the best and most interesting voyages and
          travels in all parts of the world. 17 vols., London.
PIZARRO, PEDRO.
    1917. An account of the conquest of Peru (1534). Trans. and annotated by
          P. A. Means. New York.
    1921. Relation of the discovery and conquest of the kingdoms of Peru. 2
          vols. Trans. by Philip Ainsworth Means. Documents and nar-
          ratives concerning the discovery and conquest of Latin America,
          No. 4. Cortes Soc. New York.
POGUE, JOSEPH E.
    1915. The turquois. Mem. Nat. Acad. Sci., vol. 12, pt. 2, 3d mem.
    1917. The emerald deposits of Muzo, Colombia. Trans. Amer. Inst. Min.
          Eng., vol. 55, pp. 910–933.
POND, ALONZO W.
    1937. Lost John of Mummy Ledge. Nat. Hist., Mag. Amer. Mus. Nat.
          Hist., vol. 39, pp. 176–185, March.

POPE, SAXTON T.
    1918. Yahi archery. Univ. Calif. Publ. Amer. Archaeol. and Ethnol.,
        vol. 13, No. 3, pp. 103–152, Mar. 6.
POWELL, JOHN W.
    1875. Exploration of the Colorado River of the West and its tributaries in
        1869, 1870, 1871, and 1872. Washington.
PRIEST, JOSIAH.
    1838. American antiquities and discoveries in the West. Albany.
PURCHAS, SAMUEL.
    1905–1907. Hakluytus posthumus, or Purchas his Pilgrimes. (Reprint of
        1625 ed.) 20 vols. Glasgow.
PUTNAM, FRED W.
    1886. *In* minutes of Semiannual Meeting, Apr. 28, 1886. Proc. Amer.
        Antiq. Soc., n. s., vol. 4, pp. 62–63.
RALEIGH, SIR WALTER.   *See* SCHOMBURGK, ROBERT H., EDITOR.
RAY, P. H.
    1885. Report of the International Polar Expedition to Point Barrow, Alaska.
        House Ex. Doc. 44, 48th Cong., 2d Sess.
REISS, W., and STOBEL, A.
    1880–87. The necropolis of Ancon in Peru. Berlin.
REPPLIER, AGNES.
    1929. Père Marquette, priest, pioneer, and adventurer. Garden City, N. Y.
RIBAULT, JEAN DE.
    1875. Narrative of the first voyage of Jean de Ribault · · · to make
        discoveries and found a colony in Florida, A. D. 1562. *In* French,
        Hist. Coll. La. and Fla., 2d ser., Memoirs and Narratives, 1527–
        1702. New York.
RICHARDSON, JOHN.
    1823. Geognostical observations. *In* Franklin, John, Narrative of a journey
        to the shores of the Polar Sea, in the years 1819–20, 21, and 22.
        App. 1, pp. 497–528. London.
ROBERTS, FRANK H. H., JR.
    1929. Shabik'eschee village: A late Basket Maker site in the Chaco Canyon,
        New Mexico. Bur. Amer. Ethnol. Bull. 92.
    1931. The ruins at Kiatuthlanna, eastern Arizona. Bur. Amer. Ethnol.
        Bull. 100.
ROGERS, MALCOLM J.
    1929. Report of an archaeological reconnaissance in the Mohave sink region.
        Archaeology, vol. 1, No. 1, pp. 1–13. San Diego Mus.
    1936. Archaeology, vol. 8, pp. 100–103. San Diego Mus.
RUST, HORATIO N.
    1905. The obsidian blades of California. Amer. Anthrop., n. s., vol. 7,
        pp. 688–695.
SAHAGÚN, BERNADINO DE.
    1829–1830. Historia general de las cosas de Nueva España. 3 vols. Mexico.
    1880. Histoire générale des choses de la Nouvelle Espagne. Tr. par D.
        Jourdanet. Siméon, Paris.
SALMERON, GERONIMO ZARATE.   *See* BOLTON, HERBERT L., EDITOR.
SAVILLE. MARSHALL H.
    1921. Reports on the Maya Indians of Yucatan, by Santiago Mendez [and
        others]. Edited by Marshall H. Saville. Ind. Notes and Monogr.,
        vol. 9, No. 3, Mus. Amer. Ind., Heye Foundation.
    1922. Turquoise mosaic art in ancient Mexico. New York.

SCHENCK, W. E., and DAWSON, E. J.
  1929. Archaeology of Northern San Joaquin Valley. Univ. Calif. Publ.
         Amer. Archaeol. and Ethnol., vol. 25, pp. 289–411.
SCHOMBURGK, RICHARD.
  1922. Travels in British Guiana, 1840–44. Trans. by W. E. Roth. 3 vols.,
         Georgetown, B. G.
SCHOMBURGK, ROBERT H., EDITOR.
  1848. The discovery of the · · · Empire of Guiana · · · in 1595 · · · by
         Sir W. Raleigh. Hakluyt Soc. Publ. [No. 3]. London.
SCHOOLCRAFT, HENRY R.
  1851–1857. Historical and statistical information respecting the history,
         condition, and prospects of the Indian Tribes of the United
         States. Pts. 1–6 [6 vols.], Philadelphia.
  1853. Scenes and adventures · · · in the semi-Alpine region of the Ozark
         Mountains. Philadelphia.
SCHRADER, FRANK C.; STONE, RALPH W.; and SANFORD, SAMUEL.
  1917. Useful minerals of the United States. U. S. Geol. Surv., Bull. 624.
SCHUMACHER, PAUL.
  1879. U. S. Geographical Surveys of the Territory of the United States west
         of the 100th Meridian. Vol. 7. Archaeology.
  1880. The method of manufacture of several articles by the former Indians
         of southern California. 11th Ann. Rep. Peabody Mus. Amer.
         Archaeol. and Ethnol., 1878. [Reports for 1876–1879 in 1 vol.]
         Pp. 259–264.
SELLERS, GEORGE E.
  1886. Observations on stone-chipping. Ann. Rep. Smithsonian Inst. 1885,
         pt. 1, pp. 871–891.
SHEA, JOHN GILMARY, EDITOR.
  1861. Early voyages up and down the Mississippi. Albany.
SHEPHERD, HENRY A.
  1890. The antiquities of the State of Ohio. Cincinnati.
SHETRONE, HENRY CLYDE.
  1930. The Mound-Builders. New York.
SILLIMAN, B.
  1881. Turquoise of New Mexico. Amer. Journ. Sci., 3d ser., vol. 22, pp.
         67–71, July–December.
SMITH, C. D.
  1877. Ancient mica mines in North Carolina. Ann. Rep. Smithsonian
         Inst. 1876, 441–443.
SMITH, PHILIP S., and MERTIE, J. B.
  1930. Geology and mineral resources of northwestern Alaska. U. S. Geol.
         Surv., Bull. 815.
SMYTH, LT. W., and LOWE, F.
  1836. Narrative of a journey from Lima to Para. · · · London.
SNOW, CHARLES H.
  1891. Turquoise in southwestern New Mexico. Amer. Journ. Sci., 3d ser.,
         vol. 41, pp. 511–512.
SOUTH DAKOTA STATE HISTORICAL SOCIETY.
  1902. South Dakota Historical Collections, vol. 1, appendix.
SPENCE, JAMES MUDIE.
  1878. The land of Bolivar. 2 vols. Boston.

Spier, Leslie.
    1930. Klamath ethnography. Univ. Calif. Publ. Amer. Archaeol. and
        Ethnol., vol. 30.
Spinden, Herbert J.
    1908. The Nez Percé Indians. Mem. Amer. Anthrop. Assoc., vol. 2, pt. 3.
Spruce, Richard.
    1908. Notes of a botanist on the Amazon and Andes. 2 vols. London.
Squier, E. G.
    1870. Observations on a collection of chalchihuitls from Central America.
        Ann. Lyceum Nat. Hist. N. Y., vol. 9, pp. 246–265.
Squier, E. G., and Davis, E. H.
    1847. Ancient monuments of the Mississippi Valley. Smithsonian Contr.
        Knowledge, vol. 1. June.
Stade[n], Hans.
    1874. The captivity of Hans Stade of Hesse, in A. D. 1547–55, among the
        wild tribes of eastern Brazil. Trans. by Albert Tootal. Hakluyt
        Soc. Publ. No. 51. London.
Stefánsson, Vilhjálmur.
    1913. My life with the Eskimo. New York.
    1914. Prehistoric and present commerce among the Arctic Coast Eskimo.
        Can. Geol. Surv., Mus. Bull. No. 6, Anthrop. Ser., No. 3, Ottawa.
    1919. The Stefansson-Anderson arctic expedition of the American Museum.
        Ethnological Report. Anthrop. Pap. Amer. Mus. Nat. Hist., vol.
        14, pt. 1.
    1922. Hunters of the Great North. New York.
Steward, Julian H.
    1937. Ancient caves of the Great Salt Lake region. Bur. Amer. Ethnol.
        Bull. 116.
Stoney, Lt. George M.
    1900. Naval explorations in Alaska: An account of two Naval expeditions
        to northern Alaska. U. S. Naval Institute, Annapolis, Md.
Strong, Moses, et al.
    1882. The quartzites of Barron and Chippewa Counties from the notes of
        Messrs. Strong, Sweet, Brotherton and Chamberlin. Geol. Wis.,
        Surv. 1873–79, vol. 4, pt. 5, pp. 573–581. Madison.
Sumner, William Graham.
    1907. Folkways; a study of the sociological importance of usages, manners,
        customs, mores, and morals. Boston.
Sylvester, Nathaniel Bartlett.
    1877. Historical sketches of northern New York and the Adirondack Wilder-
        ness. Troy.
Thomas, Alfred Barnaby.
    1935. After Coronado: Spanish exploration northeast of New Mexico, 1696–
        1727. Norman, Okla.
Thompson, J. Eric.
    1936. Archaeology of South America. Field Mus. Nat. Hist., Anthrop.
        Leaflet 33.
Thwaites, Reuben Gold, Editor.
    1896–1901. Jesuit Relations and allied documents. Travels and explorations
        of the Jesuit missionaries in New France, 1610–1791. 73 vols.,
        Cleveland. (See Gravier.)
    1904–1907. Early western travels. 32 vols., Cleveland.
Townsend, C. W., Editor.
    1911. Captain Cartwright and his Labrador Journal. Boston.

TUTTLE, CHARLES R.
     1885. Our North Land.   Toronto.
TYLOR, EDWARD B.
     1861. Anahuac: or Mexico and the Mexicans, ancient and modern.   London.
UHLE, MAX.
     1903. Pachacamac.   Philadelphia.
ULLOA, ANTONIO DE, and JUAN, GEORGE.
     1813. A voyage to South America . . . by Don George Juan and Don
          Antonio de Ulloa.   Trans. fr. Spanish by John Adams, Esq.   *In*
          Pinkerton's Voyages, vol. 14, pp. 313–696, London.
VAILLANT, GEORGE E.
     1930. Reconstructing the beginning of a history.   Nat. Hist., Mag. Amer.
          Mus. Nat. Hist., vol. 30, No. 6, November–December.
VANDERBURG, WILLIAM O.
     1937. Reconnaissance of mining districts in Clark County, Nev.   U. S. Bur.
          Mines, Infor. Circ. 6964.
VEATCH, A. C.
     1917. Quito to Bogota.   New York.
VÉRENDRYE, PIERRE G. DE LA.   *See* LA VÉRENDRYE.
VERRILL, A. HYATT.
     1929. Old civilizations of the New World.   Indianapolis.
VETANCURT, AGUSTIN DE.
     1870–1871. Teatro Mexicano.   T. 4, Mexico.
VON MARTIUS, CARL FRIEDERICH PHIL.
     1867. Beiträge zur Ethnographie und Sprachenkunde, zumal Brasiliens.
          2 vols.   Leipzig.
WALLACE, ALFRED R.
     1853. A narrative of travels on the Amazon and Rio Negro.   London.
WARDEN, DAVID B.
     1832–1833. Chronologie historique de l'Amerique.   10 vols., 1826–44.   (Vols.
          5–6, Historie de l'Empire Brésil, 1932–33).   Paris.
WASHINGTON, H. S.
     1922. The jade of the Tuxtla statuette.   Proc. U. S. Nat. Mus., vol. 60,
          pp. 1–12.
WEST, GEORGE A.
     1910. Pipestone quarries in Barron County.   Wis. Archeologist, vol. 9,
          pp. 31–34.
     1934. Tobacco, pipes and smoking customs of the American Indians.   Bull.
          Pub. Mus. City of Milwaukee, vol. 17, pt. 1, June 11.
WEYER, EDWARD MOFFAT.
     1929. An Aleutian burial.   Anthrop. Pap. Amer. Mus. Nat. Hist., vol. 31,
          pt. 3.
     1932. The Eskimos; their environment and folkways.   New York.
WHITE. LESLIE A.
     1932. The Acoma Indians.   47th Ann. Rep. Bur. Amer. Ethnol., 1929–30,
          pp. 17–192.
WILCOX, FRANK N.
     1934. Ohio Indian trails.   Cleveland.
WILSON, THOMAS.
     1899. Arrowpoints, spearheads, and knives of prehistoric times.   Rep.
          U. S. Nat. Mus. 1897, pt. 1, pp. 811–988.

WINCHELL, H. N.
     1884. The geology of Pipestone and Rock Counties.   Geol. and Nat. Hist.
          Surv. Minn., 1872–82; final rep., vol. 1, Geology of Minnesota,
          pp. 533–561.   Minneapolis.
WINSHIP, GEORGE PARKER.
     1896. The Coronado Expedition, 1540–42.   14th Ann. Rep. Bur. Amer.
          Ethnol., 1892–93, pt. 1, pp. 329–613.
WOOD, WILLIAM.
     1634. New England's prospect.   London.
WYMAN, LELAND C.
     1936. Navaho diagnosticians.   Amer. Anthrop., n. s., vol. 38, pp. 236–246.
ZALINSKI, EDWARD R.
     1907. Turquoise in the Burro Mountains, New Mexico.   Econ. Geol.,
          vol. 2, pp. 464–492.

LOS CERRILLOS OPEN CUT.

MINING CATLINITE PIPESTONE, MINNESOTA.

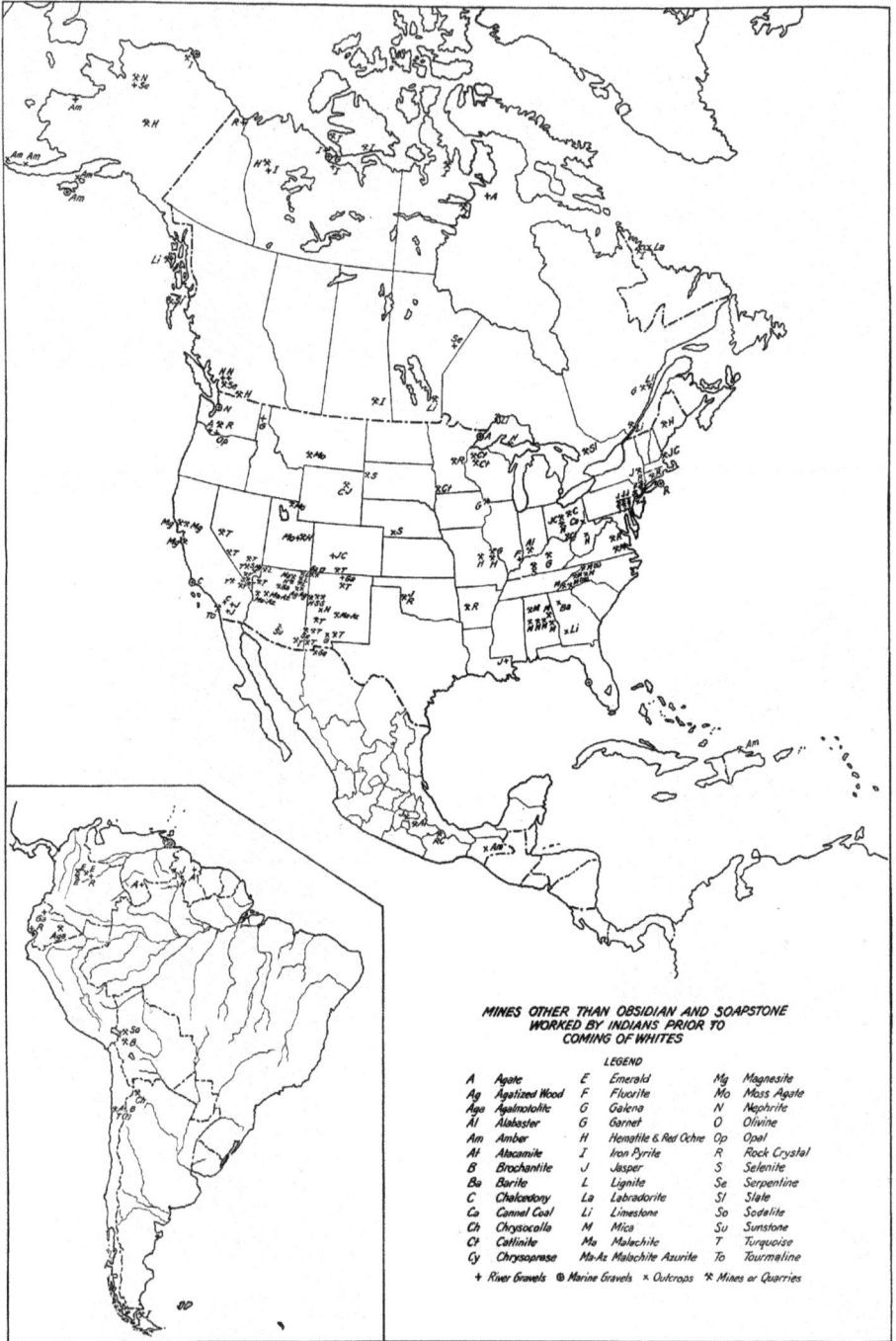

MINES OTHER THAN OBSIDIAN AND SOAPSTONE
WORKED BY INDIANS PRIOR TO
COMING OF WHITES

LEGEND

A	Agate	E	Emerald
Ag	Agatized Wood	F	Fluorite
Aga	Agalmatolite	G	Galena
Al	Alabaster	G	Garnet
Am	Amber	H	Hematite & Red Ochre
At	Atacamite	I	Iron Pyrite
B	Brochantite	J	Jasper
Ba	Barite	L	Lignite
C	Chalcedony	La	Labradorite
Ca	Cannel Coal	Li	Limestone
Ch	Chrysocolla	M	Mica
Ct	Catlinite	Ma	Malachite
Cy	Chrysoprase	Ma-Az	Malachite Azurite

Mg	Magnesite
Mo	Moss Agate
N	Nephrite
O	Olivine
Op	Opal
R	Rock Crystal
S	Selenite
Se	Serpentine
Sl	Slate
So	Sodalite
Su	Sunstone
T	Turquoise
To	Tourmaline

+ River Gravels   ⊕ Marine Gravels   × Outcrops   ⋆ Mines or Quarries

MINES OTHER THAN OBSIDIAN AND SOAPSTONE WORKED BY INDIANS PRIOR TO COMING OF WHITES.

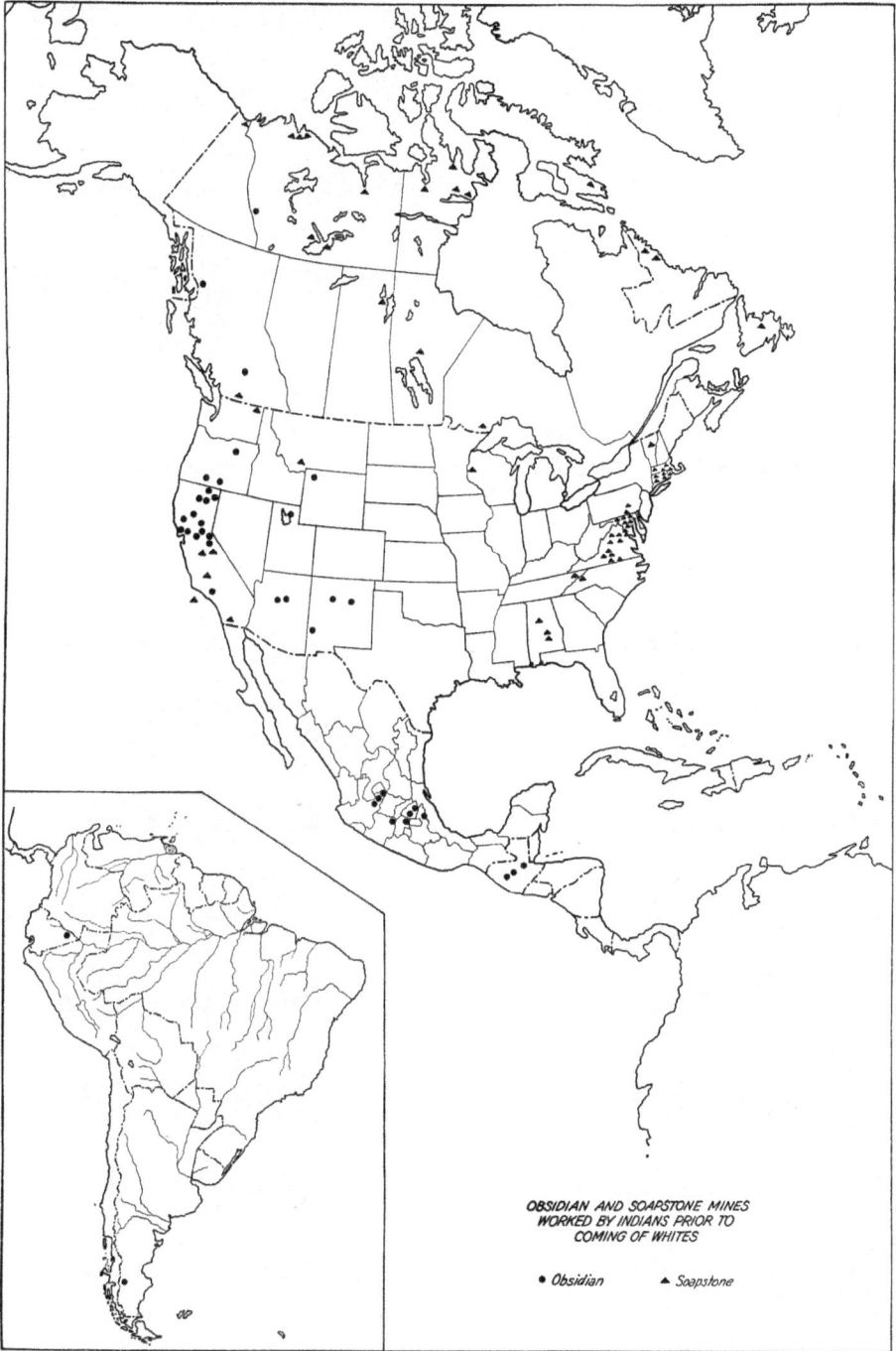

OBSIDIAN AND SOAPSTONE MINES
WORKED BY INDIANS PRIOR TO
COMING OF WHITES

• *Obsidian*   ▲ *Soapstone*

OBSIDIAN AND SOAPSTONE MINES WORKED BY INDIANS PRIOR TO COMING OF WHITES.

www.ingramcontent.com/pod-product-compliance
Lightning Source LLC
Chambersburg PA
CBHW050614210326
41521CB00008B/1246